BOSTON STUDIES IN THE PHILOSOPHY OF SCIENCE

VOLUME LV

THE LANGUAGE OF PHILOSOPHY

SYNTHESE LIBRARY

MONOGRAPHS ON EPISTEMOLOGY,

LOGIC, METHODOLOGY, PHILOSOPHY OF SCIENCE,

SOCIOLOGY OF SCIENCE AND OF KNOWLEDGE,

AND ON THE MATHEMATICAL METHODS OF

SOCIAL AND BEHAVIORAL SCIENCES

Managing Editor:

JAAKKO HINTIKKA, *Academy of Finland and Stanford University*

Editors:

ROBERT S. COHEN, *Boston University*

DONALD DAVIDSON, *University of Chicago*

GABRIËL NUCHELMANS, *University of Leyden*

WESLEY C. SALMON, *University of Arizona*

VOLUME 117

BOSTON STUDIES IN THE PHILOSOPHY OF SCIENCE

EDITED BY ROBERT S. COHEN AND MARX W. WARTOFSKY

VOLUME LV

MORRIS LAZEROWITZ

THE LANGUAGE OF PHILOSOPHY

Freud and Wittgenstein

D. REIDEL PUBLISHING COMPANY

DORDRECHT-HOLLAND / BOSTON-U.S.A.

Library of Congress Cataloging in Publication Data

Lazerowitz, Morris, 1909–
    The language of philosophy.

    (Boston studies in the philosophy of science ; v. 55)
(Synthese library ; v. 117)
    Bibliography: p.
    Includes index.
    1. Analysis (Philosophy) 2. Semantics (Philosophy)
3. Wittgenstein, Ludwig, 1889–1951. 4. Freud, Sigmund, 1856–1939.
5. Moore, George Edward, 1873–1958. 6. Contradiction. I. Title.
II. Series.
Q174.B67   vol. 55 [B808.5]   101   77–23068
ISBN 90–277–0826–6

Published by D. Reidel Publishing Company,
P.O. Box 17, Dordrecht, Holland

Sold and distributed in the U.S.A., Canada and Mexico
by D. Reidel Publishing Company, Inc.,
Lincoln Building, 160 Old Derby Street, Hingham,
Mass. 02043, U.S.A.

Printed in The Netherlands

For Alice

Nothing is more usual than for philosophers ... to engage in disputes of words, while they imagine that they are handling controversies of the deepest importance and concern.

Hume

The truth is not always probable.

Freud

EDITORIAL PREFACE

The cornerstone of the radical program of positivism was the separation of science from metaphysics. In the good old days, the solution to this demarcation problem was seen as a way of separating sheep from goats — synthetic or analytic propositions, which were candidates for truth or falsity, either on empirical or formal grounds, on the one hand; and, on the other hand, those deceptive propositions which appeared to be truth claims, but were instead either meaningless, or nonsensical, or poetical, or merely expressive — i.e. without proper cognitive content and hence not verifiable or falsifiable. In the last half century, analytic philosophy of science, linguistic analysis and logic have formulated and reformulated and reformulated again the distinction between science and metaphysics. The outcome was unexpected: metaphysics rather than retreating from the scene, shriveling up, dissolving, has instead re-emerged as an acknowledged, even if problematic component of scientific theorizing. This has made some philosophers of science happy, others miserable. For Morris Lazerowitz, it should serve as evidence for his thesis that the sort of philosophical assertions which metaphysics makes about the nature of things are not simply radically undecidable, but persist because they express a deeper and unconscious function in the mental life of philosophers (and presumably, of those scientists who find it necessary to be philosophical in this way, whether self-consciously or tacitly).

In this collection of Morris Lazerowitz's essays, Wittgenstein and Freud are the thinkers who dominate — and in a way which relates them. The themes are not new in Lazerowitz's work. In his 1955 book on *The Structure of Metaphysics*, he had already dealt with the nature of philosophical problems, and had offered his thesis that metaphysical theories are introduced as proposals for linguistic innovation; they express deep and apparently unconscious beliefs, anxieties and wishes. They are therefore not statements of problems soluble by either rational or empirical means, at least not in the ordinary sense. Rather they state 'problems' which are *dissolved*, to use Wittgenstein's term, once they .are understood. There has been much made of Wittgenstein's view (and of

Freud's roughly cognate view) that philosophical problems require 'therapy' for their solution — whether linguistic or psychoanalytic. Schopenhauer (in *The World as Will and Representation*) had already expressed such a view of solipsism as a theory, which he called 'theoretical egoism'. He said that "it was more in need of a cure than of a refutation". But Lazerowitz eschews any simple version of this view. He writes,

The suggestion here is not that a philosophical problem is a kind of aberration, but rather that it is something to understand. The equation which emerges is: understanding a philosophical problem rightly = solving the problem. No one is cured, but our understanding is enlarged. The important thing to be grasped about the nature of a philosophical problem, which makes it utterly unlike a mathematical or scientific problem, is not that understanding it is a prerequisite for its solution, but that it *is* its solution.

What Lazerowitz sets forth in this book is not simply a radical skepticism about philosophical claims as truth claims. Rather, he undertakes to show *why* such claims are made, to understand just what it is philosophers are doing with language, when they make ontological claims. In his usual acute and provocative way, Lazerowitz deals with such varied philosophical concepts as contradiction, necessity, infinity; with G. E. Moore's ontological program; with Spinoza's well-known proposition that the order and connection of ideas is the same as the order and connection of things; and throughout, with metaphilosophy, with (philosophical?) reflection on philosophy itself. The essay on Freud and Wittgenstein gives us the conceptual and critical background of Lazerowitz's own approach as well as perspicuous and striking interpretation of these thinkers. But Lazerowitz is not simply repeating, nor even merely applying their approaches. He innovates. Like J. O. Wisdom (whose psychoanalytic interpretations of philosophy, e.g. of Bishop Berkeley's thought, and whose philosophic interpretations of psychoanalysis, are related to Lazerowitz's), he finds a way to take linguistic analysis beyond the superficial reaches of mere 'clarification' or 'explication of meanings'. Indeed, in this function of what he calls 'semantic unmasking', Lazerowitz practices the high art of demystification, of dis-illusionment. The task here is to reveal that what *appears* as a truth claim, in a certain form of language, masks a deeper function of language, and a problem in the unconscious mental life of the philosopher. Feuerbach, over a century ago (in *The Essence of Christianity*), had attempted a similar task of revealing that natural human needs, wishes, modes of self-consciousness, were at the basis of their fantastic formulation in the language of religious belief and theology. "The secret of theology is anthropology", he wrote. Freud continued the Feuerbachian tradition in his essay, *The Future of an Illusion.* It is no accident that the final essay in this book is entitled 'The Passing of an Illusion.' Lazerowitz is saying, in effect, 'The secret of philosophy is the unconscious.' No easy thesis to defend, certainly; nor one which is going to find popular approval by philosophers. Even Wittgenstein, says Lazerowitz, once he recognized this consequence of Freud's theory, pushed it aside, or, in effect, sublimated it, by turning it upon philosophy itself. When Lazerowitz writes that the "fantasied omnipotence of thought is the mainspring of philosophical methodology", or that Spinoza's metaphysics was the way in which Spinoza dealt with his anxiety about any further excommunications, and that it expressed a desire for self-sufficiency, which in turn was in effect a wish to return to the security of the womb – we are bound to be disturbed.

Yet disturbance is what critical and reflective philosophy is about. And with Lazerowitz, as with Aristotle, we may "have to philosophize even in order not to philosophize." Lazerowitz's book is more than critical. It is constructive as well. He proposes a *theory* of philosophy – of what philosophers are *really* doing, when they make ontological assertions. Agree or disagree. But read on, and be provoked, stimulated, upset, enlightened by these probing and delightful essays in philosophical understanding – or in understanding philosophy.

Center for Philosophy and History of Science   ROBERT S. COHEN
Boston University   MARX W. WARTOFSKY

TABLE OF CONTENTS

FOREWORD

Anyone who works in technical, reasoned philosophy will sooner or later become aware of a number of disquieting features attaching to his discipline. Two of these must impress themselves with special force on his mind: the permanent debatability of every one of its propositions, and the absence of genuine concern in himself as well as in his colleagues over this unnatural state of affairs. The explanatory hypothesis which is developed in these pages is that behind the familiar appearance it presents, a philosophical theory is essentially an unconsciously engineered semantic deception. Disguised by the external appearance of being about what there is and about how phenomena operate are a remodeled piece of everyday nomenclature and a constellation of affectively toned fantasies which are active in the depths of the mind. The remodeled nomenclature stands between the intellectual illusion and the unconscious material, and is dynamically bound up with them: it produces the illusion that the words describe reality and it gives veiled expression to the fantasies. In a metaphor, the piece of remade terminology is a linguistic Atlas-Antaeus who holds up the philosophical firmament and in turn is held up and prevented from collapsing by the unconscious ideas to which it gives expression. Linguistic analysis, by a process which may be described as semantic unmasking, brings to light the things a philosopher does with words when he announces and argues for his theory. And an unclouded view of the linguistic structure has the effect of removing a mental blind spot and dispelling an intellectual mirage.

The hypothesis advanced here will be recognized as having its source in two recent thinkers, Sigmund Freud and Ludwig Wittgenstein. The idea that a philosophical utterance presents a piece of remapped language comes from some of Wittgenstein's later work, and it provides us with an explanation of the infinite debatability of philosophical propositions. Verbal rejection and verbal adoption are the forces at work in philosophical disputation. Preferred grammar (to use Wittgenstein's word), not fact, is in contest. The notion, deriving from Freud, that unconscious fantasies play a role in conscious thinking, and the further idea, that they play a special role in philosophical theory, helps us understand the philosopher's chronic unawareness of the nature of his activity as well as the continued fascination the theory holds for him. He transforms an unconscious conflict which he cannot resolve into a conscious problem he endlessly tries to solve. To put it in Hume's words, a philosopher 'seeks with eagerness, what forever flies [him]'. The motivation of this book is the wish to clear up the mystery to which Hume calls attention.

Conway, Massachusetts   MORRIS LAZEROWITZ
January, 1977

ACKNOWLEDGEMENTS

For permission to reprint I make grateful acknowledgement to Allen and
Unwin for Chapters III and VI; to *Critica* for Chapter I; and to *Ratio* for
Chapter V.

M.L.

METAPHILOSOPHY

When a delusion cannot be dissipated by the facts of
reality, it probably does not spring from reality.
Freud

A new field of investigation called 'metaphilosophy', which has its roots in a
number of revolutionary ideas of Ludwig Wittgenstein, is emerging, and
beginning to attract attention. By and large, however, it would seem that philosophers prefer the word to the deed, for although the word 'metaphilosophy'
has won unexpected popularity, the special investigation it was coined to
refer to has been more talked about than practiced. Nevertheless, the investigation is making itself felt, even to the point of doing some subterranean
work in the thinking of those philosophers who are careful to avoid it. Metaphilosophy is the investigation of philosophical utterances, with the special
aim of reaching a satisfactory understanding of what in their nature permits
the intractable disagreements which invariably attach to them. To an onlooker
the disagreements might well appear to have built-in undecidability, and the
assurance of philosophers who espouse rival positions to be nothing more
than a delusive state of mind.

A character in a television play remarked 'It would seem that to mortal
minds philosophical problems can never be brought to an end.' History
abundantly supports this remark, but the idea that it might be the case cannot be welcome to philosophers. Some even seem to suffer from the fear that
the investigation which attempts to discover what prevents their solution by
mortal minds will make technical, academic philosophy disappear. One
philosopher has declared that on the metaphilosophical view I have developed
over the years, metaphysics is 'the mescalin of the elite', which would make
understandable the general attitude towards the meta-investigation of philosophy. Another philosopher has coined the term 'meta-evaporate', evidently
to suggest that something is evaporated in consequence of a certain kind of
investigation. We may permit ourselves a conjecture about the determinants
which entered into the formation of this word. In his lectures Wittgenstein
sometimes said that philosophical problems have no solutions but only

dissolutions. The implication is that the clear understanding of a philosophical question removes its problem aspect. It is not a wholly idle speculation to think that the word 'meta-evaporate' was sired by the term 'metaphilosophy' and Wittgenstein's word 'dissolve', the background idea being that metaphilosophy dissolves philosophy. A flier announcing a new philosophical journal carried the following panegyric: 'Historically, philosophy has been recognized as the well-spring of wisdom and thought for a select few'. Freud, too, thought that philosophy was only for the elite of intellectuals, but he had a divided mind about it. For he vigorously resisted any attempt to introduce it into psychoanalysis and was even opposed to its use as an ornament to enhance his own subject.

The classical image of the philosopher is that of the Platonic investigator of reality in its ultimate aspects, who by the power of his thought is able to survey 'all time and all existence'. This description applies to empiricist philosophers as well as to rationalists, which is why the investigations of neither require them to leave their study. There can be no doubt that this also is the secret self-image of the linguistic analyst, the philosopher who seems farthest removed from metaphysics. Scratch the surface of the linguistic philosopher and you will find the metaphysician or the Anselmic theologian underneath. It is safe to say, quite in general, that it is the image of the omnipotent thinker which binds philosophers to their subject. Odin, whose omnipotence of thought was symbolized by the raven Hugin, paid with one of his eyes for the privilege of gazing into the well of knowledge, and one may wonder what a philosopher pays for his 'knowledge'.

The answer to this is not far to seek. Without arguing the matter here, the fact that there is no explaining philosophical disagreements on any of the usual ways of understanding the nature of philosophical theories and arguments makes it clear that the theories and arguments are misconceived. And the glaring fact that philosophers have little or no enthusiasm for looking squarely at the astonishing permanence of the disagreements shows that they have no wish to understand the nature of their activity. Freud has observed, 'If you repudiate whatever is distasteful to you, you are repeating the mechanism of a dream structure rather than understanding and mastering it.'[1] Philosophers pay for their omnipotence, or more accurately, for the spurious gratification of the wish for omnipotence, with blindness to the nature of their pronouncements, and undoubtedly the blindness radiates out and casts its shadow on more than philosophy. Like a dreamer who is able to obtain certain gratifications by keeping the meaning of his dream hidden from himself, the philosopher is able to gain satisfactions from his work by keeping it at a distance from his understanding. He pays with an inhibited intellect, in stronger words, with a weakened sense of reality, for his subjective sense of power.

Wittgenstein has remarked that philosophical propositions are not empirical, which is to say that they are not the kind of propositions to the acceptance or rejection of which either observation or experiment is relevant; and it must be granted that all the external evidence attests to the correctness of his remark. It cannot have escaped the attention of anyone that in the long history of philosophy no philosopher has attempted, nor even expressed the wish for, an experiment which would provide evidence for or against a view. The reason for this is that philosophy has no use for experiments; even those who take philosophy to be a kind of science of reality would feel the absurdity of describing it as in any way being an experimental science. In his *Scientific Thought*, C.D. Broad distinguishes between analytical philosophy, which according to him is concerned solely with the clarification of concepts, and speculative philosophy, which employs a method or methods other than analysis. In his words, the object of the latter is 'to take over the results of the various sciences, to add to them the results of the ethical and religious experiences of mankind, and then reflect upon the whole. The hope is that, by this means, we may be able to reach some general conclusions as to the nature of the universe, and as to our position and prospects in it.'[2]

The nature of the philosophical reflection on material gleaned from religion and the sciences which is to lead to general conclusions about the universe is left a mystery. It is not an analytical activity, on Broad's own account. Neither are the speculations about the universe the sort of speculations which might be confirmed or discredited by special experiments, comparable to those which have led scientists to their conclusions. The only method left open to the synoptic philosopher would seem to be observation. This is a possibility which needs to be looked into; for philosophers do sometimes give the impression of resorting to meticulous scrutiny, although it hardly needs to be remarked that they have no use for such perceptual aids as microscopes. Thus Hume appears to resort to careful observation in his investigation of the question as to whether a thing is a substance in which experienceable attributes inhere. He describes himself as carefully examining an object, such as an orange, and making an inventory of what each of his senses reveals to him. By sight he is made aware of a color and a shape, neither of which is a substance, and so on for the other senses.[3] The apparent outcome of his investigations of what his senses reveal to him is that in no case of perceiving a thing do they reveal a substance. This reminds us of A.J. Ayer's claim that all that our senses reveal to us are sense data. His claim looks like a generalization that issues from an examination of instances, an examination in which objects which are not sense data are never found. But it cannot be this; for it is not linked with the description of a theoretical counter case, of a case occurring in which something is revealed to us by our senses but is not a sense datum. The philosophical words 'is revealed by the senses but is not a sense datum' have been given no descriptive use.

It is not difficult to see that Hume's inventory is not the empirical activity it is pictured as being. He represents himself as looking for an impression of substance, which he reports he never finds. It is easily seen, however, that he could not have been looking for anything at all. For according to his own maxim that there is no idea which is not a copy of an impression, he had no idea of anything he might be described as seeking. His putative search could be no more than the imitation of a search. Hume puts himself in the case of someone who looks for an object named by a word which has not been made the name of anything. Whatever Hume's investigation was, it could not have been a search for an object. It can also be seen that his statement that colors and shapes are not substances is not the outcome of an examination of colors and shapes. He did not come to the conclusion that they are not substances by subjecting colors and shapes to a special scrutiny. The statement that colors and shapes, as well as odors, tastes, and the like, are not substances is not an inductive generalization based on an examination of instances, a generalization which could in principle be upset, for example, by a color turning up which *is* a substance. It thus seems reasonable to conclude that the statement that a color is not a substance makes an entailment-claim, in which case the sentence 'No color is a substance' would be restatable as 'Being a color entails not being a substance'.

A curious feature of the putative entailment-claim that neither a sound, taste, nor color is a substance should be noticed, as it throws the claim into an unexpected and enigmatic light. According to Hume, the conclusion to be drawn from his search for substance in the manifold of his sense contents is that we have 'no idea of substance, distinct from that of a collection of particular qualities, nor have we any other meaning when we either talk or reason concerning it'.[4] It is quite clear that Hume's supposed search for substance was not a search for a 'collection of particular qualities'; it was a search for something *in addition* to such a collection, something, moreover, he confessed he did not find. Thus, the implication of his words is that the name of what it was he was looking for, in the English language the term 'substance', has no literal sense, that it is a meaningless word like 'crul', and was not the name assigned to something for which he might look. The expression 'searches for an impression of crul', and the supposed entailment-statement that neither a color nor a shape is a substance turns out to be no more than the spurious imitation of an entailment-statement. Since, on Hume's own account, the word 'substance', as it is used by substratum philosophers like Locke and Descartes, is devoid of literal sense there can be no entailment between the meaning of a color- or shape-word, and *the meaning* of the word 'substance'. The sentence 'Being a color entails not being a substance' no more expresses an entailment-claim than does the sentence 'Being a color entails not being a crul'. Both would seem to be equally nonsensical. Nevertheless, there is an important difference between the sentences, which we might express by saying that *in some way* we do understand Hume's claim while not understanding the other, that *in some way* 'Being a color entails not being a substance' is not senseless, in the way in which the other is. Hume's statement presents us with an enigma which nevertheless we understand.

To return to Hume's description of himself as taking a careful inventory of the various contents of his sense-experience, what it shows is that a philosopher is capable of seriously misdescribing his procedure, i.e., the method he employs in his investigation. He uses, naturally and certainly without conscious guile, the language of observation and empirical search, behind which he does work of an entirely different sort. Taking Hume's description at face value, the kind of activity it refers to is *logically* different from the kind of thing he does. And in general it is clear that technical, academic philosophy which uses lines of reasoning to support its theories, is no more an observational science than it is an experimental one. Broad's distinction between analytical and speculative philosophy puts the philosopher who reaches 'general conclusions about the universe' in the position of someone who is not qualified by his special training for his speculations: he unwittingly assumes the role of amateur — who rushes in where angels fear to tread. If we look soberly at the views about the universe which cosmological philosophers like Descartes, Spinoza, and F.H. Bradley have advanced, we can easily realize that they are supported by the kind of reasoning on concepts with which we are familiar in philosophy. Wittgenstein's remark that the propositions of philosophy are not empirical makes understandable the fact that they are not supported by empirical evidence, either experimental or observational.

Now, it is natural to think that if the propositions of philosophy are not empirical, they are *a priori*, i.e., either logically necessary or logically impossible. But what is natural to think about propositions outside of philosophy may, nevertheless, not be true of propositions *in* philosophy. For philosophy is a subject that is shrouded in mystery. Until an explanation is forthcoming of the glaring but disregarded fact that one of the oldest of the intellectual disciplines is the least productive of secure results, however minor, it cannot be said that philosophical work is carried on in the open light. The fact that philosophy does not in its vast collection of propositions possess a single assertion which is uncontroversial shows philosophy to be an unnatural subject and its propositions not subject to natural and familiar classifications.

If we look with care at the *arguments* adduced in support of philosophical views, if so to speak we draw near to the arguments, we can see that they are *a priori in character*, or, at any rate, that they are *like a priori* arguments which are used to demonstrate propositions. A classical scholar remarked that Parmenides had an argument but no evidence for his view about the nature of Being.[5] We might say that an argument, which uses no experiential evidence, is the only kind of consideration which could be relevant to a *philosophical* proposition, and that empirical evidence is not logically relevant either to the support or refutation of a nonempirical proposition. An *a priori* argument may be described, in general terms, as the kind of argument which establishes, or attempts to establish, a logically necessary connection between concepts, i.e., a connection which holds under all theoretical conditions.[6] In Kantian language, a thinker who constructs an *a priori* argument does not leave the domain of concepts to obtain his result.

There is and has been for a great many years a difference of opinion regarding the nature of *a priori* reasoning. Some philosophers have maintained that it can only be dissection analysis, which consists of stating explicitly in the predicate of the resulting proposition the components which are implicit in its subject-term. The view about the nature of logically necessary propositions which is linked with this idea about *a priori* argumentation is that they are analytic: 'being logically necessary' is equated with 'being analytic'. Other philosophers maintain that not all *a priori* arguments are merely explicative and that some of them establish a necessary relation between one concept and another concept of which it is not a component. The existence of propositions which are both logically necessary and are such that their predicates (or consequents) cannot be extracted from their subjects (or antecedents) by a process of concept-dissection is thought in some quarters to make philosophy possible as an *a priori* science of reality. Philosophers who are sensitive to the actual practice of philosophy realize that it is confined to 'reflecting'[7] on concepts, in other words, that it is confined to *a priori* procedures. For the most part these philosophers rest their hope of philosophy's being informative about reality on the claim that some propositions are logically necessary and also add to our knowledge of what is referred to by the subject. A recent complaint among philosophers has been that clarity is not enough, which is to say that dissection analysis, whose function is to 'clear up'[8] concepts, conveys no information about the universe. It has to be pointed out, however, that some able philosophers have held that analytic propositions are both laws of thought and also laws of things. Regardless of this division of opinion it will be clear that an investigation which confines itself to the mere scrutiny of concepts will result in entailment-claims, either analytic or synthetic.

The word 'analysis' has a narrow use in philosophy, in which it applies to a procedure that issues only in analytic statements, and a wider use in which it applies to any procedure that issues in entailment-statements, whether tautological or not. In philosophy where it has not resulted in uncontested entailment-claims, i.e., where analysis fails to help us decide whether an entailment-claim is correct or not, it is particularly important to get a clear view of it. It has been maintained that analysis is not a procedure of defining a word but is instead a kind of examination of extra-linguistic objects, concepts or propositions, which are the meanings of expressions. Thus G.E. Moore has written 'To define a concept is to give an analysis of it; but to define a word is neither the same thing as to give an analysis of that word, nor the same thing as to give an analysis of any concept'.[9] The impression created by these words, and

others to a similar effect, is that philosophical analysis is not in any way an
examination of language and that language is external to conducting an
analysis: angels could practice analysis without the help of language. Wittgenstein's reported suggestion that we ask for the use of an expression, not for
its meaning, gives rise to a different idea of analysis. This is that analysis is a
special kind of investigation of the use terms have in a language. The nature
of philosophical analysis is bound up with the kind of claims it results in, i.e.,
putative entailment-claims, and whether it is the investigation of verbal usage
or of extra-linguistic entities can best be determined by getting clear about
the nature of entailment-statements, or of logically necessary propositions.

Without entering into refinements, entailment-statements may be said to
divide into two kinds: those which may be called 'identity-entailments',
i.e., those whose consequents are identical either with their antecedents or
with a component of their antecedents, and those whose consequents are not
related to their antecedents in this way. The former are held by many philosophers to have no factural content, that is, to have no use to convey information about things or occurrences, etc. Wittgenstein has epitomized this
feature of tautologies by remarking in the *Tractatus* (4.461) that 'I know
nothing about the weather when I know that it is either raining or not raining',
or what comes to the same thing, when I know that if it is not raining, then it
is not raining. Some philosophers might oppose this by saying that this
tautology is about the weather and tells us what the weather *must be*. We
may allow this, but we shall then have to say that a proposition which tells
us what the weather must, logically, be fails to tell us what the weather *is*.
The proposition remains true no matter what the weather is like and so tells
us nothing about what the weather actually is. In Wittgenstein's words
(4.462) 'tautologies are not pictures of reality'; and this is because their
truth-values are not determined by what the world happens to be like. A lady
won control over her wayward dog by saying to it, 'Tucker, come here or
don't come here!' Her dog could not fail to do what she told it to do, which
gave her subjective gratification. But Tucker could not fail to do as he was
told because what he was told was not a command: he was not ordered by
her words to do anything. Similarly a tautological statement, or an identity-
entailment, has no use to convey information about things or occurrences,
because being true under all conditions it says nothing about which condition obtains.

Some philosophers, e.g., Leibniz, have held that all propositions are really analytic. Others, amongst them Wittgenstein,[10] have held that there are no analytic propositions. And still others, e.g., John Stuart Mill, have held that there are no necessary propositions of any kind. As is known, philosophers are also in disagreement over whether the class of necessary propositions is identical with the class of identity-entailments; and those who agree among themselves that there are non-identity entailments, i.e., synthetic *a priori* propositions, are by no means in agreement over which necessary propositions are synthetic. Despite this puzzling multiplicity of conflicting opinions (which no one really expects to be resolved so long as philosophy is an active enterprise), most philosophers have the idea that the possibility of there being an *a priori* investigation of reality depends on there being propositions which are both synthetic and logically necessary. But philosophers who rest their hope on the existence of such propositions rest it on a feeble reed, one that is too fragile to provide a solid perch for the owl of Minerva. The reason for this is that the consideration which shows that tautologies have no use to give factual information about things shows the same thing about synthetic *a priori* propositions. In general, a proposition which is characterized by logical necessity is *prevented* from being about the existence and behavior of things by the fact that it is necessary.

A proposition which is true by necessity is one whose truth-value is invariant under all theoretical conditions, and this is because its truth-value is not conditioned by what there is. Wittgenstein has said: 'A tautology has no truth-conditions, since it is unconditionally true.'[11] Those words, suitably modified, apply to every *a priori* truth; it is unconditionally true, and thus is not related to conditions, actual or theoretical, which *make* or *would make* it true. What prevents a tautology from having factual content is just the fact that it is logically necessary or is true *independently* of conditions, and this also prevents a synthetic *a priori* proposition from having factual content. A proposition which is true independently of what the world is like, is true no matter what it is like, and thus has no use to convey information about what there is or about what there is not. According to a Kantian claim, the words 'Every change has a cause', unlike the words 'Every effected change has a cause', express a non-identity entailment, namely, *being a change* entails *having a cause*. But if the words do in fact express an entailment, then they say nothing more about conditions under which changes occur than do the words 'Every caused change has a cause'. Philosophers who, like Broad and A.C. Ewing, hold the so-called entailment view of causation would seem to be holding a view which is not about how events are related to each other or about a condition under which changes occur.

Many philosophers who give up the notion that *a priori* truths can be about things (or, it would be better to say, give up this notion with part of their mind) go over to the position that they are verbal, about the actual use of terminology. One version of this view is that they 'simply record our determination to use words in a certain fashion';[12] another version is that they are true by definition or by virtue of the meanings certain words have; and still another version is that 'they are purely about the use of the expressions they connect'.[13] On this view, the analytic proposition that an effect must have a cause is about the use of the words 'effect' and 'cause' which it connects, and it is made true by the use these words have; and the putative synthetic *a priori* proposition that a change must have a cause is about the use of the terms 'is a change' and 'has a cause'. On the conventionalist idea of logical necessity, it will be clear that a synthetic *a priori* proposition no more than one that is analytic is about things. It needs to be remarked, however, that so-called linguistic philosophers have been charged with thinking that facts about things can be inferred from the study of verbal usage, and that the truth-values of philosophical theories can be determined by the study of the language in which the theories happen to be formulated.[14]

Conventionalism is not free from objections, which are well known. One of them is the following. A proposition that is about the use of terminology is empirical: usage could be different from what it actually is, which is to say that a true verbal proposition could in principle be false, and of course conversely. Thus, if the *proposition* expressed by the sentence 'An effect must have a cause' were about the use of the words 'effect' and 'cause' and how they function with respect to each other in the English language, it would be empirical. But if it is empirical, i.e., has one of two possible truth-values, which remain its *possible* truth-values independently of the truth-value it actually has, it cannot be *a priori*, i.e., be such that its actual truth-value is its only theoretically possible truth-value. A proposition cannot be both logically necessary and also verbal. Just as a proposition is prevented from being about things by being *a priori*, so it is prevented from being about words by being *a priori*. A logically necessary proposition is neither about things nor about words.

A third view as to what necessary propositions are about has to be noted, even if only briefly here. This is that their subject-matter is a special part of all that there is, the domain of abstract objects. A nonempirical proposition issues from a scrutiny of concepts and their connections, and a correct analysis results in an *a priori* proposition which is a truth about concepts. A sentence which expresses a proposition that is true by *a priori* necessity expresses an entailment that holds between abstract entities. Without going into Platonic metaphysics, we can best see *what* we know when we know that a proposition is *a priori* true by considering *sentences* which express them, and comparing them (a) with sentences which express propositions about things (or about ordinary, nonabstract objects) and (b) with related sentences which express propositions about verbal usage.

Consider the following sentences.

> All gamps are large hand umbrellas.
> All gamps, without the theoretical possibility of an exception, are large hand umbrellas.
> Being a gamp entails being a large hand umbrella.

A person who did not know the word 'gamp', if he were informed that the word was in the dictionary and that the sentence expressed a true proposition, might think it expressed an empirical generalization, like that expressed by the sentence 'All Watusi are very tall'. But this information would not be enough to tell him the meaning of the word 'gamp'. The information that the second sentence expresses a true proposition would, however, be enough to tell him what the word means, and this is also the case with regard to the third sentence. Knowing that the last two sentences express a true proposition is the same as knowing that the first sentence states a logically necessary proposition. Hence what a person learns who is informed of the fact that 'All gamps are large hand umbrellas' states an *a priori* proposition, is a verbal fact to the effect that a term has a certain meaning. What we know when we know that a sentence expresses a logically necessary truth, and know this in virtue of understanding the terms occurring in the sentence, are verbal facts about the use of terminology. But the sentence does not *express* what it is that we know in knowing that it expresses a necessary proposition: the proposition is not verbal, although knowing it to be true requires nothing more than knowing facts of usage. A person can hardly be blamed if he feels like the peasant in the fable who is told by the satyr that he blows on his fingers to warm them and on his soup to cool it. It would seem to require a semantic satyr to say, on the one hand, that to know that a proposition is logically necessary all that is required is knowledge of rules for the use of words, and on the other hand, that the proposition is nevertheless not verbal!

What makes it difficult to explain the nature of *a priori* necessity and creates the impression that we are blowing both verbally hot and verbally cold, is the form of speech in which a sentence must be cast in order to express a necessary proposition. In a metaphor, this mode of speech gives a sentence for an *a priori* necessity the two faces of Janus, each of which belies the other. Its verbal face and its nonverbal, ontological face can be made visible by comparing the sentence

A gamp is a large hand umbrella

with the sentences

The word 'gamp' means large hand umbrella.
A gamp is an awkward thing to carry.

The first sentence is related to the second in a way that is easier to see than to describe. The second sentence, we might say, *states* what it is that we know in knowing that the first sentence makes a true *a priori* claim; but the first does not *state* what the second states. The first sentence conveys, without expressing, what the second openly expresses. To put the matter in oracular language, the sentence 'The word "gamp" means large hand umbrella' *reveals* what the nonverbal sentence *conceals*: the explicit verbal content of the one is the concealed verbal content of the other. To put the difference in terms of the familiar distinction between use and mention, the second sentence mentions a term which occurs without mention in the first. In this respect, the second sentence has a feature in common with the third sentence, 'A gamp is an awkward thing to carry', which mentions no word that occurs in it. But the two sentences differ in an important respect, and this is a difference which may be said to be truly invisible: the third sentence *uses* the word 'gamp' to refer to something, actual or imagined, whereas the word occurs in the first sentence without being used to refer to anything. Both are in the ontological form of speech, the form of speech that is used to make statements about the existence and nature of things. Nevertheless, unlike the third sentence, the first does not *use* words to refer to things. In a figure, the resemblance between the two sentences is only skin deep; beneath its linguistic surface the sentence for the necessary proposition has its content in common with the verbal sentence. The sentence, 'A gamp is a large hand umbrella', presents a verbal fact in the ontological mode of speech.

In sum, a sentence which expresses a necessary proposition is in the form of speech in which language is used to describe things and occurrences, but is nevertheless a sentence that does not use words to make a declaration about things. It is this feature, its ontological form, which, despite the verbal content of the sentence, is capable of creating the delusive idea that language is being used to express a theory about the world. The analysis of *concepts* turns out, accordingly, to be the explication of verbal usage conducted in the ontological idiom; and, again, it is the form of speech which is responsible for the idea that philosophical analysis is a method of learning basic facts about things 'by the mere operation of thought',[15] by mental penetration into the meanings of expressions. Thus, the proof that the diagonal of a square is incommensurable with its side, far from looking like the explication of the rules governing the use of terminology in a regularized, exact language, has the appearance of being a kind of unfolding of contents hidden in meanings which when brought to the surface reveal a mystifying fact about ideal objects. Undoubtedly it was the wish to dispel this appearance and to make us aware of the linguistic substructure which supports the appearance that made Wittgenstein characterize a logically necessary proposition as a 'rule of grammar'.

It will be remembered that Wittgenstein said that philosophical propositions are not empirical, which makes it natural to infer that he thought them to be *a priori*. Some philosophers, for one reason or another, have attempted to cast doubt on the validity of the distinction between empirical and *a priori* propositions. This cannot be gone into here, but we can permit ourselves one jaundiced remark: this is that behind the nominal rejection of the distinction, philosophers continue to use the distinction, which they mark by a different terminology. An air of a verbal game being played surrounds the rejection. The unavoidable impression made on an observer who finds himself puzzled by the game is that some philosophers are made unhappy by conventional nomenclature and wish to change it. In ordinary life most of us seem to be made less unhappy by the invented term 'senior citizens' than by the ordinary expression 'old people'. Wittgenstein said that a philosopher rejects an expression with the thought that he is refuting a theory, and this appears to apply to philosophers who 'refute' the validity of the empirical-*a priori* distinction.

To return to the point that philosophical propositions are not empirical. Seeing that they are not makes it natural to proceed to the idea that they are *a priori*. This idea fits in which the analytical procedure by which they are arrived at. But as has already been noted, what is a natural inference outside of philosophy may, nevertheless, not be a correct inference *in* philosophy. By way of parenthetical observation, it should be noticed that if the views of philosophy are *a priori*, then they are certainly not cosmic propositions, propositions about what the universe is or is not like. The sentences which express them convey only information about the actual use of terminology in a language; they are 'rules of grammar', in Wittgenstein's way of speaking, formulated in the nonverbal mode of speech. It is hardly necessary to say that philosophers would deny that their questions were merely verbal or that they are centrally concerned to explicate usage. And in a way their disavowal is correct: philosophy makes use of analytical lexicography but does not reduce to it. A philosophical sentence like 'Chance is nothing but concealed and secret cause'[16] is linked with the explication of the actual use of the words 'change' and 'cause', but does not serve to convey information or misinformation about actual usage. Construed as expressing a necessary proposition it would be equivalent to the entailment-sentence 'Being a chance occurrence entails having an unknown cause'. Part of the verbal claim obliquely presented by the sentence is that 'uncaused chance occurrence' has no application to occurrences, actual or theoretical, in other words, that the phrase has no descriptive content. Some so-called ordinary-language philosophers, and also Wittgenstein at times, would say that instead of expressing a necessary proposition a philosopher who asserts the sentence is in a disguised way misrepresenting actual usage. An alternative construction which can now be placed on the philosophical sentence is that it neither expresses an *a priori* truth nor misrepresents usage, but rather is a disguised way of announcing an artificially retailored use of the word 'cause', a stretched use in which it applies to all occurrences, chance as well as non-chance.

Philosophers are not made happy by the thought that their views have only verbal content. The idea of an *a priori* science of the universe has great appeal, and undoubtedly receives its support from the unrelinquished wish for cosmic clairvoyance. Sober reflection, which seems to have no chastening effect on unconscious wishes, shows that if a claim is about things it is not *a priori* and if it is *a priori* it is not about things. Locke said that mathematics is both demonstrative and instructive. It is not to be denied either that mathematics is demonstrative or that it is instructive; but if we resist Platonic metaphysics, the reality which emerges is that in respect of being demonstrative mathematics in an indirect way is about the use of expressions it connects. Thus, for example, learning by means of a calculation that what the sentence '$13 = 187 - 174$' says is true comes to knowing a connection between the use of the term '13' and the expression '$187 - 174$'. To put it very roughly, mathematics explicates, and thus is instructive of, the interrelations of the rules governing the functioning of its symbols, which is not to imply that it has no application to certain kinds of problems about things.

One consideration that goes against the view that philosophical sentences which fail to express *a priori* truths misrepresent the actual use of terminology is the resistance to correction by those who pronounce them. A person who states that $34 \times 7 = 228$ will see his mistake when it is pointed out to him. But a philosopher who asserts that a chance occurrence has a secret cause has an unaccountable myopia to his mistake, despite his knowing what everyone else knows about the actual use of 'chance' and 'cause'. Even if he acknowledges the mistake, there is no guarantee that later he will not return to it. It is not necessary to be a scholar of the history of philosophy to learn that a refuted philosophical theory is a phoenix which springs into life from its own ashes over and over again.

The metaphilosophical view which construes a philosophical sentence as embodying an academically retailored piece of terminology presents an explanation of what makes it possible for a philosopher to remain fixated to his 'mistake', or if he gives it up, makes it possible for him to return to it. Put simply and without elaboration, the re-editing of terminology, unlike the misrepresentation of verbal usage, is not open to correction. One involves a mistake, the other does not. A philosophical 'view' is a grammatical creation, which like a daydream can be enjoyed alongside the world of sober fact with which we have to live. To hark back to Wittgenstein's important observation that a philosopher rejects a form of words while fancying himself to be upsetting a proposition about things: we might instead say that the philosopher is covertly changing language under the illusion that he is revealing to us the contents of the cosmos.

We are now in a position to understand Hume's baffling statement that a color is not a substance, baffling because in conjunction with showing to his own satisfaction that we have no idea of substance, he states that a color is not a substance. The sentence 'We have no idea of substance' is such that if it expressed a true proposition it would be the case that the word 'substance' stands for no idea, or that it is a word without meaning; and obviously, anyone who thought that it expressed a true proposition would treat 'substance' as he treats a nonsense word. If it expressed a true proposition, the sentence 'A color is not a substance' would not express a proposition about what a color is not; it would be literally senseless. There is a strong temptation to say that Hume just failed to see that his sentence was, on his own showing, nonsensical; but another explanation is now available which also has to be considered. This is that he was, to use Wittgenstein's expression, playing a language game with the word 'substance', now doing one thing with it, now another, for the dramatic effect the game produces.

To put the matter overbriefly and without argument, Hume's claim that 'we have no idea of substance, distinct from that of a collection of particular qualities' has *grammatical import.* It comes to saying that substantives are words that stand for *qualities*, or what is the same thing, that nouns are really adjectives.[17] If we do not look on Hume's statement as a concealed misdescription of the grammar in actual use, we come to the idea that it presents an ontologically expressed grammatical rearrangement of parts of speech, the classification of nouns with adjectives. His philosophical view that there is no substance distinct from a cluster of qualities puts forward a grammatically reconstituted language in which words that normally count as substantives are to count as adjectives. This rearranged grammar is projected onto the everyday noun-adjective structure of language, and gives rise for some people to the magical illusion that a discovery has been made about the nature of things. To others it gives rise to the illusion that the real structure of our language, its depth grammar, is being revealed. The statement that a color, or a shape, is not a substance connects up with the fact that words for colors, shapes, tastes, etc. have a double grammatical classification: they count in our grammar both as adjectives and as nouns. What Hume's use of the statement is, the work he makes it do, now comes into plain view. The sentence 'Colors and shapes, etc., are not substances' is used to introduce a contracted grammar for words denoting colors, shapes, etc., one in which they are shorn of their noun use in the language.

NOTES

[1] S. Freud, *A General Introduction to Psychoanalysis*, p. 130.
[2] C.D. Broad, *Scientific Thought*, p. 20.
[3] *A Treatise of Human Nature*, Book I, Part I, Section VI.
[4] Ibid.
[5] Benjamin Farrington, *Greek Science. Its Meaning for Us*, p. 49.
[6] An *a priori* consideration may, of course, show concepts to be independent of each other, e.g., the concepts *cow* and *having two stomachs*.
[7] C.D. Broad's term.
[8] C.D. Broad's term.
[9] 'A Reply to My Critics', *The Philosophy of G.E. Moore*, The Library of Living Philosophers, Vol. IV (ed. P.A. Schilpp), p. 665. See also p. 661.
[10] *Notebooks 1914-1916*, p. 21.
[11] *Tractatus Logico-Philosophicus*, 4.461. D.F. Pears and B.F. McGuinness translation.
[12] A.J. Ayer's way of putting it.
[13] John Wisdom, *Philosophy and Psychoanalysis*, p. 63n.
[14] See C.D. Broad, 'Philosophy and "Common-Sense"', in A. Ambrose and M. Lazerowitz (eds.), *G.E. Moore. Essays in Retrospect*, esp. p. 203.
[15] Hume's expression.
[16] Hume attributes this view to other philosophers. *A Treatise of Human Nature*, Book I, Part III, Section IV.
[17] Interestingly enough, Hume makes a similar remark about the problem of personal identity: '. . . all the nice and subtle questions concerning personal identity can never possibly be decided, and are to be regarded rather as grammatical than as philosophical difficulties.' *A Treatise of Human Nature*, Book I, Part IV, Section VI.

CHAPTER II

FREUD AND WITTGENSTEIN

In [some] cases the repressed pleasure in looking is changed into an unproductive desire for knowledge which is not applied to real events.

Karl Abraham

I

Sigmund Freud, the creator of psychoanalysis, made what is perhaps the most remarkable breakthrough in the history of science: he was the scientific discoverer of the unconscious, and he devised a method for bringing its contents, the thoughts and wishes buried in it, into conscious view. Phenomena pointing to the existence of an active region of the mind which is detached from consciousness, such as post-hypnotic suggestion and everyday errors into which we naturally read hidden motivation, had been known long before Freud, but they failed to excite enough scientific curiosity to initiate any serious investigation. To use an expression from Kant, before Freud the world was in a state of dogmatic slumber with regard to such phenomena. Some thinkers, notably Plato, Nietzsche, and Schopenhauer, had a small number of perceptions into the workings of the unconscious, but their revelations had only the value of literary conversation pieces, and were treated as such. By contrast, Freud's discoveries aroused the most hostile and stormy opposition imaginable. The reactions against them resembled the outraged reactions of people whose deepest secrets have been bared to the world. Even nowadays, when the medical value of psychoanalysis is widely recognized and there is general realization that its findings have become permanent additions to science, Freudian theory continues to live under a cloud.[1] It is made to suffer the kind of punishment the Pennsylvania Dutch community metes out to a member who violates a taboo. By and large, with some exceptions, psychoanalysis is surrounded by hostile silence and misrepresentation.

There is no question but that psychoanalysis is looked on by many people as being in some way a threat[2], against which they must protect themselves. It is important, therefore, to try to bring into the open what in psychoanalytic teachings is so emotionally disturbing. One thesis, to which philosophers particularly have expressed opposition, is that part of the mind is unconscious and its contents inaccessible to us. We apply the pronoun 'I' to the conscious part of our mind, with which we identify ourselves, and the idea of an active part of the mind which lies beyond the reach of its scrutiny and control presents an unappealing picture of ourselves. Psychoanalysis represents us as strangers in our own mental home and as subject to alien impulses and thoughts. Small wonder that it arouses opposition: if the shoe fits, it pinches and cannot be worn. Furthermore, special psychoanalytic claims, such as those regarding infantile sexuality and the Oediapal wishes, turn civilized mankind, in general, against psychoanalysis. Thus, in recent years the claim has been made that psychoanalysis may have been useful in the past but may now be discarded because of present day sexual permissiveness. It is not difficult to see that this is a spurious reason, since, to mention one thing, incest and parricide can hardly be said to be socially permitted or approved. The important thing is to come to terms, at least at an intellectual level, with the 'outrageous' ideas of psychoanalysis, if we are to form a detached judgment of psychoanalytic theory.

The rejection of the unconscious needs to be looked at with special care, because, as it turns out, the rejection can be either of two utterly different kinds, which nevertheless look alike. One is a factually empty, *verbal*, rejection of the unconscious, while the other is a matter of fact denial of its existence. One is the philosophical suppression of the term 'unconscious mental process', while the other is the denial that the phenomenon to which the term is used to refer actually exists. Freud spoke of philosophers 'for whom "conscious" and "mental" are identical'[3]. He did not, however, distinguish in a clear-cut way the philosopher's identification of the conscious and the mental from that of the ordinary person who has an antipathy to psychoanalysis and rejects the claim that there is an unconscious part of the mind. The philosopher's identification is based on a rejection of terminology, whereas the ordinary person's identification is based on a rejection of a claimed phenomenon. In not recognizing this with sufficient clarity Freud to a degree was taken in by the semantic legerdemain of the philosopher. When a philosopher like Descartes or Locke declares that consciousness is an invariable accompaniment of mental processes, or that there is no such thing as an unconscious mental occurrence, he is not, contrary to the appearances, saying something that is at all like saying that there are no poltergeists. For the philosopher's identification of 'conscious' and 'mental' is such that they cannot even in theory part company. In his way of speaking, consciousness is a logically inseparable accompaniment of mental processes, which is to say, in his language the term 'conscious' applies to whatever 'mental' applies to. His *philosophical* use of 'conscious' departs from its everyday use, but not, to make a parenthetical observation, as the result of a mistake about the actual use of terminology, which he knows as well as anyone else.

That Freud in a way recognized the distinction between the verbal claim that the terms 'mental' and 'conscious' are interchangeable and the factual claim that mental phenomena are in every instance conscious appears in the following statement: 'It seems like an empty wrangle over words to argue whether mental life is to be regarded as co-extensive with consciousness or whether it may be said to stretch beyond this limit, and yet I can assure you that acceptance of unconscious mental processes represents a decisive step towards a new orientation in the world and in science.'[4] But an 'empty wrangle over words', that is, a dispute which is about words rather than things, may not be empty of unconscious motivation, which needs to be brought to light. It is obvious that the stipulation that 'mental life' (and like terms) not be stretched beyond the range of application of the word 'conscious', or that its application be confined to the range of mental phenomena covered by 'conscious', would, if adopted, prevent the expressing of psychoanalytic theory. Its effect would be to silence psychoanalysts.

The philosophical denial that there could be a mental occurrence, a wish or a thought, from which consciousness is absent may easily be seen to be cut off from any sort of evidence which would tend to make a philosopher give up his claim or make him allow that it lends a degree of probability to the position he opposes. Freud stated that 'the study of pathogenic repressions and of other phenomena compelled psychoanalysis to take the concept of the unconscious seriously'[5], but the findings of such a study could not possibly go against the philosopher's position regarding the nature of mental phenomena. This is because his claim is not open to theoretical falsification. His claim, we might say, is immune to refutation by evidence; and the reason for this is that his claim prevents him, linguistically, from saying what confuting evidence would be like. Wittgenstein perceived this when he remarked that philosophers who do not allow the possibility of unconscious thoughts rule themselves out from allowing the possibility of conscious thoughts: 'They state their case wrongly when they say: "There can only be conscious thoughts and no unconscious ones." For if they don't wish to talk of "unconscious thought" they should not use the phrase "conscious thought", either.' [6] The implication of Wittgenstein's remark is that the philosophical rejection of the notion of an unconscious thought amounts to the rejection of the *expression* 'unconscious thought'. That is, in the philosophical way of speaking the word 'unconscious' has no use to characterize thoughts, or to set some thoughts off from others. By stating that a philosopher who rejects the expression 'unconscious thought' should not use the term 'conscious thought', Wittgenstein quite clearly implies that the philosopher can no longer use 'conscious' to characterize any thoughts. The view that all thoughts are conscious thoughts, which appears to present a reassuring factual claim about thoughts, thus shrinks into the view that all thoughts are thoughts, which obviously is impregnable against any sort of evidence.

It is clear that the tautological statement 'All thoughts are thoughts' is barren of factual content: it has no more use to convey information about thoughts than 'If there are horses, then there are horses' has to convey information about the existence of horses. To use Wittgenstein's expression, it 'says nothing' about thoughts; and there is some temptation, to which Wittgenstein may have succumbed, to go on and maintain that such a statement says nothing whatever.[7] There is less temptation, however, to think that the *philosophical* statement 'All thoughts are conscious thoughts' (once it is seen that it, too, makes no factual declaration) says nothing; instead there is some temptation to construe it as being a verbal statement about actual usage. Thus, according to conventionalism, it says nothing regarding a property of thoughts, but it does say something about the correct application of the word 'conscious': namely, that the rule for its use dictates its application to whatever counts as a thought. In sum, the expression 'unconscious thought' has no use to describe a thought and the descriptive use of 'conscious thought' is the same as that of the word 'thought'.

Construed as stating a matter of verbal usage, the plain implication of the philosophical statement is that in relation to the word 'thought' (and also in relation to 'wish', 'impulse', 'motive', etc.) 'conscious' is a semantically useless word. If in fact 'unconscious' had no use to characterize thoughts, its antithetical term would have no such use either, and would be *idle* in the language. It is a fact, of course, that the word is not descriptively idle, that it does have a use to characterize thoughts, and, thus, that 'unconscious' also has such a use. Philosophers who reject the unconscious know this perfectly well. Their grasp of language is not inferior to that of anyone else. There is only one way of understanding the philosophical statement which avoids our having to think that they both know usage and also persistently have a mistaken idea about it. This is to construe it, not as making a claim about the actual use of 'conscious' (and of 'unconscious') but as embodying a changed use which is more pleasing to the philosopher than the familiar one. It would seem that the philosopher plays a game with the words 'unconscious' and 'conscious': he suppresses the first, prevents it from applying to anything mental, and retains the second by making it apply to everything mental. In doing this he makes the word 'conscious' descriptively functionless. He does this in such a way, however, that it does not appear to be idle in his philosophical vocabulary. He creates the impression that he is disclosing a feature of all mental occurrences, while underneath he is merely playing a game with words. The main gain to him is that he is able to delude himself into the comforting belief that he has upset a disturbing view about the nature of the mind. He removes from existence an unacceptable reality by the manoeuvre of exorcising the word 'unconscious' out of the language. We may guess that the delusion is supported by magic thinking, still alive in his mind, which identifies a thing with its name. He fantasies a word out of the language, and in this way fancies himself to be destroying what the word stands for.

A person who *philosophically* rejects the unconscious does not need to consider empirical evidence either for or against the psychoanalytic claim. He satisfies himself of the correctness of his own position by a procedure in which the evaluation of evidence plays no sort of role. But anyone who opposes Freudian theory, not by erasing a word from the language but by facing up to the theory itself, is obliged to acquaint himself with evidence cited in the literature. He cannot be present in the psychoanalyst's consultation room, but he can read case studies. A person who wishes to make a serious examination of the general and special findings of psychoanalysis is not in the predicament of a sceptic in the time of Copernicus, before Galileo had made the first rudimentary telescope; he need do not more than consult the literature. Here only a piece of pre-psychoanalytic evidence, which Freud reports in his *Autobiographical Study*, will be given. One thing which led him to take seriously the notion that powerful mental processes could be at work in the mind while hidden from consciousness was the study of pathogenic repression. The following passage, which is about the famous patient of 'the talking cure' case, deserves careful reading.

The patient had been a young girl of unusual education and gifts, who had fallen ill while nursing her father, of whom she was devotedly fond. When Breuer took over her case it presented a variegated picture of paralyses with contractions, inhibitions and states of mental confusion. A chance observation showed her physician that she could be relieved of these clouded states of consciousness if she was induced to express in words the affective phantasy by which she was at the moment dominated. From this discovery, Breuer arrived at a new method of treatment. He put her into a deep hypnosis and made her tell him each time what it was that was oppressing her mind. After the attacks of depressive confusion had been overcome in this way, he employed the same procedure for removing her inhibitions and physical disorders. In her waking state the girl could no more describe than other patients how her symptoms had arisen, and she could discover no link between them and any experiences of her life. In hypnosis she immediately revealed the missing connection. It turned out that all her symptoms went back to moving events which she had experienced while nursing her father; that is to say, her symptoms had a meaning and were residues or reminiscences of those emotional situations. It turned out in most instances that there had been some thought or impulse which she had had to suppress while she was by her father's sick bed, and that, in place of it, as a substitute for it, the symptom had afterwards appeared. But as a rule the symptom was not the precipitate of a single such 'traumatic' scene, but the result of a summation of a number of similar situations. When the patient recalled a situation of the kind in a hallucinatory way under hypnosis and carried through to its conclusion, with a free expression of emotion, the mental act which she had originally suppressed, the symptom was abolished and did not return.[8]

Classical Greek philosophy early discovered that nature works by unseen bodies, and the fundamental thesis of psychoanalysis is that the mind works by affectively charged invisible complexes of ideas. This thesis is held to apply to the normal, healthy person as well as to those people who are psychologically unwell. With the help of the method Freud devised for investigating mental functioning, psychoanalysis makes possible the understanding not only of pathological phenomena but also of social institutions, artistic reactions, moral systems, religion – and it helps clear up the enigma that is technical philosophy. Consider briefly one application of psychoanalytic findings to a pair of puzzling social-religious institutions widely found in primitive societies, totemism and exogamy. To make a parenthetical observation, a more regressed and less liberal form of these institutions appears to exist in the Catholic church. This is suggested by the convents and monasteries; nuns are the brides of Christ and monks are permitted no women at all. Clinical experience led Freud to the discovery of the so-called Oedipus Complex, which is given dramatic expression in Sophocles' tragedy, *Oedipus*

Rex. In the myth, from which Sophocles fashioned his play, the oracle at Delphi tells Oedipus that he will kill his father and marry his mother. This comes to be. In a chance encounter he slays Laius, his father, and later, in ignorance of who she is, marries his mother, Jocasta, and takes his father's place both as husband and king. The myth reveals the pair of instinctual wishes which are rejected with horror and fall under repression: parricide and incest.

The application of the Oedipus cluster of ideas to totemism and to exogamy brought into the light of understanding not only the meaning of the separate institutions but also the functional relation between them. An important criterion of scientific explanation is, roughly, that a theory regarding a group of phenomena, apparently occurring together only accidentally, is a possible explanation of the phenomena, if the theory brings out a functional connection between them. Freud's clinical experience showed him the way to such an explanation. His experience made it clear that the dread of incest, behind which was a powerful drive to commit it and against which elaborate precautions had to be erected, was linked to the male head of the family. The wish to remove the father is the counterpart of the forbidden longing for the mother: they are the two interacting components of the Oedipus complex. Freud's idea was that totemic worship which so often goes along with the taboo of incest is internally bound up with it: the totem represents the forbidding yet loved father, who must be protected against injury by his rivals for the mother.

Freud was led by his discoveries about the nature of dreams, in part at least, to the idea that psychoanalysis was an instrument not only for the study and cure of psychological diseases, but also for the study of normal mental phenomena, and of human creations such as totemism, agriculture, and works of art. He wrote:

Previously psychoanalysis had only been concerned with solving pathological phenomena and in order to explain them it had often been driven into making assumptions whose comprehensiveness was out of all proportion to the importance of the actual material under consideration. But when it came to dreams, it was no longer dealing with a pathological symptom, but with a phenomenon of normal mental life which might occur in any healthy person. If dreams turned out to be constructed like symptoms, if their explanation required the same assumptions – the repression of impulses, substitute-formation, compromise formation, the dividing of the conscious and the unconscious into various psychical systems – then psychoanalysis was no longer a subsidiary science in the field of psycho-pathology, it was rather the foundation for a new and deeper science of the mind which would be equally indispensable for the understanding of the normal. Its postulates and findings could be carried over to the other regions of mental happenings; a path lay open to it that led far afield, into spheres of universal interest.[9]

The unconscious, which is the underworld into which rejected wishes and thoughts are cast, surrounds the conscious life of man and apparently influences and colors it at every point. A repressed wish is not extinguished; it remains alive and constantly presses for satisfaction, sometimes with greater strength, sometimes with lesser strength, depending on the circumstances. It is permitted to enter into the conscious part of the mind, and receive indirect, substitutive gratification, only after being disguised and made unrecognizable, as it is when it enters into a dream. The disguised return of a wish may be pathological; it may return in the form of a private symptom or an obsession. Or it can be normal and shared with others. The psychological odyssey of the poet and the artist, whose work counts not only as being healthy and normal but also as a valuable contribution to civilization, is described by Freud in the following words:

The realm of imagination was evidently a 'sanctuary' made during the painful transition from the pleasure principle to the reality principle in order to provide a substitute for the gratification of instincts which had to be given up in real life. The artist, like the neurotic, had withdrawn from an unsatisfying reality into this world of the imagination; but, unlike the neurotic, he knew how to find a way back from it and once more to get a firm foothold in reality. His creations, works of art, were the imaginary gratifications of unconscious wishes, just as dreams are; and like them they were in the nature of compromises, since they too were forced to avoid any open conflict with the forces of repression. But they differed from the asocial, narcissistic products of dreaming in that they were calculated to arouse interest in other people and were able to evoke and to gratify the same unconscious wishes in them too.[10]

These words also describe the inner life of the philosopher, who is of special interest to us. He, like the poet, wins success and the respect of the intellectual world with an 'invented dream' which he can share with others. Philosophy truly is an enigma to the understanding, for it makes hardly anyone uneasy despite its not being able to hold up a single established proposition, however minor or trivial. Philosophers and their cultural admirers continue to cling to the agreeable illusion that philosophy is a science of reality. It is not to be expected that philosophers will wish to investigate their 'science' in order to discover why in its long history it has been unproductive of results. And it must be said that psychoanalysts have so far been of no help, even though they have shown the greatest interest in scientific methodology. But this is not to imply that psychoanalysis can throw no light on philosophy. Quite the contrary. Wittgenstein has shown the way to a clear understanding of the linguistic structure of the philosophical theories, which brings into plain view the reason why they can remain in permanent debate. He has enabled us to get an inside look into the structure of a deceptive semantic illusion, and psychoanalysis is able to tell us why the illusion holds such fascination for us and is so difficult to give up. Just as many people have felt the need to blind themselves to Freud's discoveries about the workings of the mind, so philosophers have felt the need to push out of consciousness Wittgenstein's deeper thoughts about philosophy. Anyone who opens his mind to these thoughts runs the risk of seeing the ground on which so many philosophical edifices have been erected sink before his eyes, and with it the entire philosophical metropolis that has been more than two thousand years in the building. This is a risk too great to take. There is no Lloyd's of London to insure them against its happening.

II

Wittgenstein's thought, after his *Notebooks, 1914-1916* (which are largely metaphysical), falls into three major periods. In the first period he worked on his *Tractatus Logico-Philosophicus*, where he was concerned primarily with conventional philosophical problems. These he looked at through the somewhat unconventional spectacles of logical positivism.[11] His main object apparently was to throw into doubt the intelligibility of metaphysics, and, following Bertrand Russell, to give logical respectability to the remainder of philosophy. It needs to be remarked that a good deal of metaphysics hides behind his rejection of metaphysics.

The second period, which marks an iconoclastic departure not only from the *Tractatus* but from other conventional ways of looking at philosophy, falls into a small number of years after his return to Cambridge in 1929. It was at this time that he dictated material which is now published under the title *The Blue and Brown Books*. G.E. Moore, Alice Ambrose, and John Wisdom attended his lectures and to a greater or lesser degree fell under his intellectual influence. The years after 1936 represent a partial return to the pre-*Blue Book, Tractatus* period. His *Philosophical Investigations*, written in these years, is given over in part to what amounts to a conventional critique of doctrines in the *Tractatus*.

Different views are current about the connection between these three periods. Professor G.H. von Wright, the successor to Wittgenstein's chair in Cambridge and one of his literary executors, has said that the second period is a puzzling discontinuity in Wittgenstein's thought, that it is something which really does not belong. In his own words, 'I myself find it difficult to fit *The Blue Book* into the development of Wittgenstein's thought.'[12] The suggestion of his words is that the period is not representative of the *real* Wittgenstein. One gathers that it is the thirteenth floor at which the elevator need not stop. Other of Wittgenstein's later followers consider the philosophy he did in this period his 'false philosophy'; it is the floor at which the elevator should not stop. The idea appears to be that Wittgenstein, being human, was capable of error; it is an aberrated period in his intellectual odyssey and should be isolated and left to wither away. It is hard not to think of these disciples as intellectual morticians who come to praise him in order to bury his heresies. There is still a third view about the connections between the periods. Some well-known philosphers maintain that there is no discontinuity in his development, that his continuous work is traditional epistemology and metaphysics. They deny that the second period exists. These philosophers are true lotus eaters, who by self-induced somnambulism erase from their minds Wittgenstein's disquieting perceptions into the subterranean workings of philosophy.

In my opinion Wittgenstein's middle period represents a fundamental break-through in a discipline which has eluded the reach of understanding for an astonishing period of time. Philosophy has for ages been a sphinx with a captivating voice. The glaring fact that technical, reasoned philosophy has not a single uncontested result remains as unnoticed as infantile sexuality, and is a riddle for which conventional philosophy provides no answer.[13] The work of Wittgenstein's middle period shows the way to an understanding of this odd state of affairs. His explanation of the nature of philosophical utterances may not be entirely correct, but without it we would remain completely in the dark.

It may be unkind but, nevertheless, true to say that too many philosophers act on the assumption that what is emotionally unpalatable is either false or does not exist. It needs to be pointed out that Wittgenstein's explanation *must* be unappealing to philosophers, since the fact on which it throws light is one that philosophers prefer to ignore rather than face up to. The statements from Wittgenstein's writings to be given shortly will make clear why a philosopher would wish to ignore one side of his thought. They bring out of the dark an activity that is intriguing and absorbing only if carried on in the dark. The risk that improving our understanding of technical philosophy may *dissolve* it will be accepted only by thinkers who place a high value on understanding. Those who prefer to cling to a valued illusion stand in little danger of losing it.

Before quoting a number of his remarks and drawing conclusions from them it is important to call attention to his change of attitude to psychoanalysis. For the insights he had into the working of academic philosophy are like the perceptions a psychoanalyst has into the forces which produce a neurosis, a dream, a reverie, a surrealistic painting, and a fairy tale. The central motive of the period when he dictated *The Blue and Brown Brooks* was the linguistic unmasking of philosophical utterances, utterances which parade as statements about the inner nature of things, about space and that which lies beyond the bounds of space, about time, causation, and so on. In a Socratic metaphor, the implication of some of Wittgenstein's later insights is that philosophical statements are semantic wind-eggs which are represented as having conscious ideational content.

When Wittgenstein first learned about psychoanalysis and read Freud he was filled with admiration and respect. He remarked about Freud, 'Here is someone who has something to say'.[14] Later he turned against psychoanalysis and rejected it as a harmful mythology. A remarkable thing, however, took place, which might be described as the hidden, displaced return of the rejected.[15] He imbued his philosophical talk with a kind of psychoanalytical atmosphere. It is as if, for him, philosophy had become a linguistic illness from the burden of which people needed to be relieved, and this could only be done by laying bare the illusion-creating tricks that were being unconsciously played with language. The impression that reading the later Wittgenstein makes on one is that he had, without being aware of it, become the psychoanalyst of philosophy, the background formula perhaps being: I don't need analysis, philosophers need it; I am the psychoanalyst, philosophy is the illness from which philosophers need to be cured by analyzing what they are doing with language.[16] It is perhaps within the bounds of reasonable speculation to think that Wittgenstein would never have had his remarkable insights into philosophy, if the need for analysis had not been deflected away from himself and projected onto philosophy. A number of philosophers seem to have divined his role with respect to philosophy when they described him as a *therapeutic* positivist. And indeed he explicitly stated that his 'treatment of a philosophical question is like the treatment of an illness.[17] One of his ideas was that 'the sickness of philosophical problems can get cured only through a changed mode of thought and of life'.[18] It is clear from these and other remarks which he imbedded in his later writings that he thought of philosophy as a neurotic aberration which called for treatment.

It would seem that Wittgenstein reacted to his own insights into the workings of philosophy in the way in which philosophers have frequently reacted to the account I have given of the nature of philosophy, an account which is merely a development of Wittgenstein's thought. At a symposium meeting organized by Professor Sidney Hook some years ago in New York, the well-known Harvard philosopher, Raphael Demos, protested that I implied that philosophers were sick (his word). D.A. Drennen, in a book of readings which included his own commentaries,[19] stated that according to my view the difference between lunatics and philosophers is that philosophers are not institutionalized. D.D. Raphael reacted with less emotion and more sobriety to the view when he observed that it represented metaphysics as 'the mescalin of the elite'.[20] In so characterizing it he seems to have come somewhat close to a distinction psychoanalysts make between therapeutic or medical analysis and so-called 'applied' analysis. The purpose of the first is to cure a person of a neurotic illness, or to lessen its severity. The purpose of the second is to add to our knowledge of a phenomenon, for example, a work of art or a primitive tribal practice like totemism, by laying bare the hidden meaning it has for us. These two aims of psychoanalysis are of course not mutually exclusive but neither are they identical, and confusing the two leads to the idea that a normal activity, like dreaming or writing a novel, is a manifestation of a psychological illness. It is this confusion which made Wittgenstein, according to reports, speak of 'dissolving' philosophical problems, after the analogy of removing a symptom. Nevertheless, he in some way recognized the difference when he spoke of eliminating the problem aspect of a philosophical question, or to put it differently, of removing its puzzlement. The suggestion here is not that a philosophical problem is a kind of aberration, but rather that it is something to understand. The equation which emerges is: understanding a philosophical problem rightly = solving the problem. No one is cured, but our understanding is enlarged. The important thing to be grasped about the nature of a philosophical problem, which makes it utterly unlike a mathematical or a scientific problem, is not that understanding it is a prerequisite for its solution but that it *is* its solution.

The following remarks about philosophy throw it into a new light. They represent an extraordinary intellectual break-through[21] in a discipline which has endured with only superficial changes for an astonishing length of time.

A philosophical problem has the form: 'I don't know my way about'. (*Philosophical Investigations*, p. 49)

The philosopher's use of language is like an engine idling, not when it is doing work. (Ibid., p. 51)

The man who says 'only my pain is real', doesn't mean to say that he has found out by the common criteria – the criteria, i.e., which give our words their common meanings – that the others who said they had pains were cheating. But what he rebels against is the use of *this* expression in connection with *these* criteria. That is, he objects to using this word in the particular way in which it is commonly used: On the other hand, he is not aware that he is objecting to a convention. (*The Blue Book*, p. 57)

And it is particularly difficult to discover that an assertion which a metaphysician makes expresses discontentment with our grammar when the words of this assertion can also be used to state a fact of experience. (Ibid., pp. 56–7)

The fallacy we want to avoid is this: when we reject some form of symbolism, we're inclined to look at it as though we'd rejected a proposition as false. . . This confusion pervades all philosophy. It's the same confusion that considers a philosophical problem as though such a problem concerned a fact of the world instead of a matter of expression. (The Yellow Book[22])

These remarks speak for themselves so clearly that it is hardly necessary to comment on them. Their implication is that a philosopher changes language in one way or another under the illusion that he is expressing a proposition about what there is (or is not) and about the nature of what there is. The further implication is that the revised piece of language is *semantically idle*, i.e., that it has no actual use to communicate information, yet creates the illusion of expressing a speculation about the world.

III

An example taken from classical philosophy will illustrate Wittgenstein's claim. Heraclitus maintained that everything constantly flows or changes, that nothing remains the same. You cannot step into the same river twice nor sit on the same bench twice, because there is no such thing as the same river or the same bench − or even the same you. There is the apocryphal tale that a debtor of Heraclitus refused to make repayment on the grounds that the present Heraclitus was not the Heraclitus from whom he borrowed the money and that he was not the original borrower. In linguistic terms, this 'view' amounts to withholding from things the application of the phrase 'remains the same', while retaining in the language the antithetical term 'changes'. A philosopher who declares that nothing remains the same for any length of time, however short, that everything flows or is in a state of continuous change, gives the impression of putting forward a factual claim about things encountered in everyday experience. To all appearances he is rejecting a common belief about things like iron anvils and granite mountains and is replacing it with a proposition about their real state. And there can be no doubt that the philosopher fancies himself to be a kind of scientist who upsets a superstition and puts in its place a reasoned truth about things. But Wittgenstein tells us that the philosopher is confused about his own activity: he thinks himself to be an investigator of reality and to have made a revolutionary discovery, while all that he is doing is rejecting an ordinary expression, banishing it from the language. Instead of investigating the nature of what there is, he plays a game with words: he rules 'remains unchanged' out of usage, while artificially retaining 'is undergoing change', but does this in a mode of speech which creates the false idea that he is making a pronouncement about things, rather than gerrymandering with terminology.

The philosopher suppresses one of a pair of antithetical terms while retaining the other.[23] Parmenides rejected the terms 'motion' and 'change' while artificially keeping in the language their antitheses. Heraclitus does the reverse: he suppresses 'remains the same'. But doing this has the semantic effect of robbing the antithesis with which it is linked of the use it has in the language. Without its antithesis 'remains the same', the phrase 'undergoes change' no longer serves to distinguish between things and, to use Wittgenstein's word, becomes linguistically 'idle'. The meaning of the phrase 'thing which undergoes change' vanishes into the meaning of the word 'thing', so that in the sentence 'All things are things that undergo change', the term 'undergoes change' loses its use and becomes semantically functionless. The meaning of the sentence contracts into that of the empty sentence 'Everything is a thing'. It is not difficult to see that the *point* of banishing 'remains the same' while artificially retaining 'undergoes change' is to create the deceptive illusion that the sentence 'Everything changes, nothing remains the same' expresses a theory about the nature of things.

The illusion the Heraclitean pronouncement creates has a certain vivacity, but nevertheless does not have sufficient strength of its own to continue to survive scrutiny without collapsing. The correct conclusion to come to is that it receives unconscious support from a fantasy to which the pronouncement gives subterranean expression. Kant said that we discover in nature what we ourselves have put there, and Freud, the first explorer of the unconscious, tells us that we project inner processes into the outer world, our projections sometimes taking the form of scientific speculations. Without stretching the imagination, Heraclitus' theory that the universe is a conflict of opposites under the control of justice can be recognized as reading, via a projection, an inner state of affairs into the outer world. His view that everything flows also, undoubtedly, derives the major part of its charge from a cluster of unconscious fantasies that are given expression by the utterance. What one of these fantasies is may be conjectured by considering the Greek from which the word 'diarrhea' derives. It would seem that the view that everything is in flux or that everything constantly changes is the concealed expression of an anal fantasy projected onto things. To sum up, the philosophical sentence 'Everything flows' creates the illusion that it is used to express a cosmic theory about things. Semantically, the sentence introduces an artificially stretched use of 'flows' in which the word applies to everything to which 'thing' applies, and underneath it has a use to express an unconscious fantasy.[24] This helps us understand what Wittgenstein may have meant by 'the bewitchment of our intelligence by means of language'.[25]

Another view may be considered briefly. Bertrand Russell has said that solipsism can neither be shown to be false nor yet be adopted as a proposition on which one can act. Taken at face value, the words 'I do not know that anyone else exists' or the words 'I alone exist' express propositions which no one in his right mind would dream of acting on. Even an avowed solipsist behaves like anyone else; he does not act like a somnambulist nor like someone who rejects what his eyes and ears tell him about the reality of other people. He greets others and responds to their greetings like anyone else. The solipsist, in his philosophical moments, may talk like a man who is out of his senses, but he never behaves like one. The differences between him and a lunatic is not that he is not institutionalized. The difference is that he does not need to be. Only some of his talk is strange, although it has to be said immediately that not even his solipsistic talk is considered strange or in any way odd in philosophy. Philosophy seems to be a sanctuary in which apparently aberrated talk is accepted as reasoned, scientific discourse. Wittgenstein has shown us the way to the window through which we may get a clear view of the nature of this sanctuary.

Mrs Ladd-Franklin, in a well-known letter, wrote Bertrand Russell that she was a solipsist, a position she found so satisfying as to recommend it to others. Apart from its comical side, her letter would seem to show a remarkable blindness to an inconsistency: the inconsistency in a sentence like 'Dear Mr Russell, I am writing to tell you that I alone am real'. The blindness is too much to be accepted, although not to accept it is not to impute sham blindness. And if there is no blindness, neither is there an inconsistency. The following passage from *The Blue Book* helps us understand why there is no inconsistency and no blindness. It should be pointed out immediately that removing the apparent blindness requires our recognizing something else that is an actual blindness — one which is imposed on us by an illusion.

Now the man whom we call a solipsist and who says that only his own experiences are real, does not thereby disagree with us about any practical question of fact, he does not say that we are simulating when we complain of pains, he pities us as much as anyone else, and at the same time he wishes to restrict the use of the epithet 'real' to what we should call his experiences; and perhaps he doesn't want to call our experiences 'experiences' at all (again without disagreeing with us about any question of fact). For he would say that it was *inconceivable* that experiences other than his own were real. . . I needn't say that in order to avoid confusion he had in this case better not use the word 'real' as opposed to 'simulated' at all; which just means that we shall have to provide for the distinction 'real'/'simulated' in some other way. The solipsist who says 'only I feel real pain', 'only I really see (or hear)' is not stating an opinion; and that's why he is so sure of what he says. He is irresistibly tempted to use a certain form of expression; but we must yet find *why* he is.[26]

The strange talk of the solipsist, whether he declares that he does not really know that anyone else exists or that he alone is real, on Wittgenstein's understanding of it is not the talk it seems on the surface to be; it is not about what is known or about who exists and who does not exist. In a supposed imaginary conversation Wittgenstein has with a philosopher he turned to a third person and explains, 'He is not mad. We are just philosophizing.'[27] We may say a like thing of the solipsist: he is *only* philosophizing. Without embarrassment either to himself or to us, he can philosophically say to our very faces: 'Only I really perceive, feel anger, have thoughts; I alone am real.' This is because his words are not used to make a declaration regarding the existence or nonexistence of anyone. Wittgenstein represents the solipsist as giving expression to a linguistic wish, the wish to 'restrict the use of the epithet "real" to what we should call his experiences'. This understanding of what he is doing with terminology brings a bright light to one of the darkest corners of philosophy. It helps us get clear on how it is that, without intellectual dishonesty, he can hold his view while sympathizing with us when we complain of pain and on why we do not think his view to be the symptom of a psychological malady. If, for whatever reason, the solipsist uses the sentence 'I alone am real' or 'Only my experiences are real', to introduce a re-edition of the word 'real' which confines its application to the solipsist, we can understand both why he 'does not disagree with us about any practical question of fact' and also why his 'view' can be the subject of intractable and endless disagreement. He is so sure of what he says because he is not 'stating an opinion'; instead he is presenting a terminological change which he prefers. His philosophical opponent can be equally sure because he is not stating a counter-opinion about a matter of fact but is merely rejecting a terminological change.

Despite semantic appearances to the contrary, the word 'real' is not, in the solipsistic way of speaking, opposed to 'simulated', and Wittgenstein's recommendation to the solipsist is that, to avoid confusion and to prevent misunderstanding, he should not, in the expression of his view, 'use the word "real" as opposed to "simulated" at all'. However, if the solipsist followed Wittgenstein's suggestion, and did not use the word 'real', or its equivalents, in the expression of his theory, his theory would evaporate; and the same thing would happen, if in some way he explicitly marked the fact that the word in his pronouncement did not have its ordinary use. It would seem that Wittgenstein had the idea that the solipsist wishes to introduce a terminological change in the *actual* use of everyday language, which differs from philosophical language in not being 'like an engine idling'. But philosophers are not language reformers, as their continued unresolved disputes show. And as Wittgenstein himself points out, if the present distinction between 'real' and 'simulated' were obliterated by adopting a change in the everyday use of 'real', we should 'have to provide for the distinction "real"/"simulated" in some other way' − which implies that from a practical point of view the change is linguistically pointless.

We reach an understanding of the solipsist's *wish* to contract the application of 'real', if we keep in mind the fact that his sentence 'I alone am real' gives the appearance of making a factual claim, an appearance to which he is dupe and which holds him in bondage. What becomes clear then is that the philosophical use of 'real' is not introduced for the sake of a putative practical advantage but rather is introduced for the sake of the illusion it brings into existence. The sentence 'I alone am real', or the sentence 'Only I really have experiences', creates the vivid and forceful, even if delusive, impression of stating an experiential proposition, and we have to think that *its special work is to produce this impression*. We may say that in the sentence, which presents in the ontological form of speech an academically gerrymandered word, the term 'real' does not have a use to convey factual information. Instead, it has a metaphysical job, which is to help create the intellectual illusion that a theory is being announced.

Wittgenstein represents the man who says ' "only I feel real pain", "only I really see (or hear)" ' as being 'irresistibly tempted to use a certain form of expression', and urges us to search for the reason why. One reason for his being attracted to his contrived use of 'real' undoubtedly is that it creates a wished-for illusion. We can also discern the special egoistic gain the philosophical view has for him, if we realize that in its actual use the word 'real' means not only *genuine* and *exists*, but also *important*. The solipsist unconsciously uses his sentence to bolster his ego. Bertrand Russell's story about the letter he received from Mrs Ladd-Franklin in which she expressed surprise that others were not solipsists has its amusing side, but it makes plain that she found the doctrine highly gratifying. Each of us can have the consolation of solipsism without deprivation to anyone else. The solipsistic sentence 'I alone am real' gives its philosophical user an 'ego monopoly' which in no way is in conflict with that of any other user of the sentence. For the ego monopoly is that of infantile self-love, which recognizes no distinction between oneself and things that are not part of oneself. The sentence is also used to give hidden expression to a somewhat later stage in our mental development, when we recognized a world of things other than ourselves in which we were the center. In this stage we were like the god of Aristotle around whom the world revolves. When we reflect on Aristotle's theological picture of the world, the idea which comes to mind is that a piece of psychological auto-biography has been projected into the cosmos.

The solipsistic view, at least the one discussed by Wittgenstein, is a structure in which three components can be discerned. One component is an academi-cally contracted use of the word 'real' which dictates its application exclusively to whatever the first person pronoun 'I' is used to refer to. A second component is the delusive appearance that the solipsistic sentence states a fact-claiming theory, an appearance which is brought about by the nonverbal mode of speech in which the sentence is formulated. And a third consists of an un-conscious fantasy of narcissistic self-aggrandizement, which is given expression by the word 'real' in the philosophical sentence.

One further view should be examined, as its underlying content appears to be ubiquitous in philosophical thinking. Kant described the 'new method of thought' which he introduced into the philosophical investigation of nature in the following way: '. . . we can know *a priori* of things only what we ourselves put into them.'[28] One thing which, according to him, we know *a priori*, is that the world operates in conformity with the principle that every change has a cause. We can be certain of this, not because of evidence provided by the natural sciences, but because we ourselves have introduced causal regularity into the world. In one place he declares that '. . . the highest legislation of nature must lie in ourselves,' to which he adds that 'we must not seek the universal laws in nature by means of experience, but conversely must seek nature as to its universal conformity to law, in the conditions of the possi-bility of experience which lies in our sensibility and in our understanding'.[29] In another place he declares that 'the order and regularity in the appearances which we entitle *nature*, we ourselves introduce. We could never find them in the appearances, had not we ourselves, or the nature of our mind, originally set them there'.[30] Kant allows that it may seem strange to assert that 'The understanding does not derive its laws (*a priori*) from, but prescribes them to, nature',[31] but he has no doubt of the truth of his 'bold proposition'.

Kant was certainly correct in describing his new method of thought as both bold and strange, but it is doubtful whether he had more than a glimmer of how very strange it was. Looked at in its abstractly stated form its oddness barely shows through, which makes it possible to present it as a bold innovation.

But if we translate into the concrete Kant's proposition that the laws which scientists think they have discovered *in* nature are really laws put into nature by our minds, it becomes strange, indeed, so strange that it is too much to suppose that anyone could at the conscious level of his mind possibly hold it. If, however, we go on to examine Kant's philosophical proposition through the lens of metaphilosophy, both its strangeness and its boldness fade away and what remains (we might say, its naked remainder) is not likely to attract or hold the interest of anyone.

If we take at face value the sentence 'Our minds prescribe to nature the laws science discovers there' and pin down its ostensible meaning to concrete implications, the result is astonishing, to say the least. Consider Oersted's law that an electrical current in a wire creates a magnetic field around it, or the law that the application of heat to a gas expands it. The implication of the ostensible theory with regard to these laws is that when there is an electrical current in a wire *our mind* produces a magnetic field around it, and that when heat is applied to a gas *our mind* makes it expand. The occasionalists held the curious view that on the occasion of a mind willing a certain bodily action God intervenes and brings about the appropriate behaviour, and Kant (apparently without full realization) seems to have held an equally odd view. The occasionalists give us the picture of God as a cosmic exchange operator; and Kant gives us the picture of a mind as a world-titan who causes the physical changes in the world which their invariant physical antecedents delusively appear to cause. There is no reason to think that Kant would be willing to reformulate Oersted's law as stating that when electricity is present our mind produces a magnetic field. Nor can we think that he would accept what seems to be a plain implication of his view, that before there were minds there were no laws of nature and that there will be none after minds pass out of existence. The fact that Kant would not accept such propositions while nevertheless holding his view shows, not that he fails to see that they were consequences of it; it shows, rather, that his view does not imply them.

If we reflect on the putative claim that our mind puts into nature the laws we find there, we can see that it does not imply that as a mere matter of fact a so-called law of nature is not a physical law, e.g., that it just is not the case, although it *could* conceivably be, that *heat* causes gas to expand and that *electricity* produces a magnetic field. Rather, it implies that it is *logically impossible* for there to be a physical law in nature. For anyone who would deny that the laws described in physics books are *physical* laws of nature would rule himself out from being able to say what a physical law of nature would be like. From this we may infer that *as he is using language*, to say that the understanding does not derive its laws from nature but puts them into it is equivalent to saying that it is logically impossible for there to be a law in nature that the understanding does not put there. The implication is that it is logically impossible for there to be a physical law, or a concomitance of occurrences, which is physically determined. But to assert that it is logically impossible for there to be a physical law, as against a regularity brought about by a mind, is in an *oblique way* to make a declaration about the use of the term 'physical law', oblique in that the declaration does not expresssly *mention* the term. What is being stated in an indirect form of language is that the term has no application, or that it has no characterizing use.

Looked on as making a claim about the actual use of an expression in a language, it is flagrantly false. It goes against linguistic reality so flagrantly that if we pause to reflect on it we cannot think it a mere mistake. A possible alternative explanation, which demands investigation, is that the philosophical sentence 'It is logically impossible for there to be a physical law', rather than making a mistaken verbal claim, presents a revised use of the term 'physical law', a use in which it is *stripped* of its everyday application. On this construction, Kant's theory that the laws we discover in nature our minds have put there, if we bring it into line with the construction placed on the implied claim that there can be no physical laws, turns out to be nothing more strange or bold than the introduction of revised terminology: a stretched use of 'law of the mind' in which it applies to what is normally referred to by 'law of nature' and a contracted use of the term 'physical law' in which it has zero application. Put in terms of the word 'cause' the language revision amounts to a contracted use of 'physical cause', which bars its application to causes, and a stretched use of 'psychological cause', which stipulates the application of 'has a psychological cause' to every occurrence.

It is plain that no sort of scientific or practical gain attaches to the Kantian re-editing of terminology. Some gain results to the language of geometry from the paradoxical definition of the word 'point' as a circle with zero radius. And some gain is made for the language of physics by the definition (which is not in agreement with everyday usage) that the word 'work' means the expenditure of energy. The gain from Kant's philosophical alteration of nomenclature seems to be the production of an intellectual illusion on which philosophers place a high value, but one that is too fragile to survive a close look. The illusion, which is colored by megalomaniac grandeur, is able to continue to exist only so long as it is kept at a distance from those who enjoy it. It collapses as soon as it is subjected to Moore's technique of translating a general theory into the concrete. No philosopher, however steeped in metaphysics his thinking may be, can at the conscious level of his mind believe that the planets are kept in their orbits by our minds or that the rate of acceleration of bodies in free fall is causally determined by our understanding. The explanation of the remarkable durability of the illusion created by the Kantian game with words is that philosophers have no wish to look with care at their creation. The absence of such a wish in thinkers who profess to be scientifically oriented indicates the presence of a *resistance* against closer looking. There is no other explanation.

We may conjecture that the reasons for a resistance which has been powerful enough to protect an illusion for such a considerable number of years are more than just to sustain the illusion, although it has to be admitted that the ego-investment of the philosopher is great. It is not hard to imagine the blow his self-esteem would suffer were he to learn that, instead of the cosmic scientist he fancies himself to be, he cuts the unheroic figure of a semantic wizard of Oz. Nevertheless, there is more to the matter than the linguistically contrived illusion which floats on the surface. Acccording to one psychoanalyst modern art, in many instances, is an expression of a narcissistic retreat from the real world; the Kantian illusion also may be viewed as a narcissistic retreat from reality, a regression in the service of the ego to an early stage in the development of its sense of reality. We all pass through a psychological stage in which no distinction is made between oneself and the external world, when the self is identified with the sum total of things. In this stage it is as if our wishes are omnipotent and create their own fulfilment; it is a stage which might be described as one of fantasied cosmic self-sufficiency. When this infantile delusion begins to be undermined by frustrations and delayed satisfactions, and later by a sense of helplessness, as is inevitable even in the most favored cases, our feelings of omnipotence go through various transformations. One of these is the transference of omnipotence to those who care for us, our parents, etc. Later there is a transference to formulas and practices, which are invested with magical powers. Thus, according to the Old Testament God said,

'Let there be light', and there was light. The thought which must come to
everyone's mind is that this represents a piece of narcissistically 'corrected'
autobiography: it harks back to the time when we were helplessly cast out of
the watery darkness into overwhelming and frightening light. Later we win
mastery over the memory of that first upheaval by a narcissistic correction.
In our unconscious we create what originally we were subjected to. The
Stoics maintained that we achieve freedom when we will that to happen
which inevitably will happen, and the Genesis invention, which is the projection of our fantasy, tells us that we willed the light into which we once
were abruptly thrown, that we created it with the magical formula 'Let
there be. . .'.

    It is not beyond the bounds of the psychological realities to think that a
metaphysical scientist who writes a book bearing the title *Mind and the
World-Order* is motivated by an idea in his unconscious, the idea that he rules
all that happens in the universe by the power of his mind. And it is not
beyond the bounds of psychological reality to think that Kant's words,
'We discover in things only what we ourselves have put there', represent a
sophisticated instance of progression, the construction of a semantic structure,
which makes possible a regression in the service of the ego. The interpretation
which with some plausibility may be placed on them is that they express a
touched-up picture of the world, one which answers to the deepest wishes
within us. The philosophical claim that we put into things what we find in
them represents an attempt to heal the breach between ourselves and the
world, and pictures a return to the original unity in which whatever we perceive is part of our self. The claim that we prescribe to nature the laws
governing it also presents a corrected picture of the world. In our unconscious
the world, which early in our development goes its own way and becomes
indifferent to our needs and wishes, is subdued and made agreeable with the
fantasy that our wishes control things. There is the story that King Canute
commanded the incoming tide to stop. A philosopher wins in his unconscious what a Canute cannot win in reality, and he does this with the help of
a piece of trumped-up language, a piece of semantic 'flimflam', to use Kant's
word.

NOTES

[1] For the attitude of philosophers to psychoanalysis, see especially Sidney Hook (ed.), *Psychoanalysis, Scientific Method, and Philosophy.*

[2] Some literary people have expresssed the conviction that the analysis of a poem, for example, would debase it and thus deprive them of an aesthetic pleasure.

[3] *An Autobiographical Study*, p. 55.

[4] *A General Introduction to Psychoanalysis*, p. 23.

[5] Ibid.

[6] *The Blue Book*, p. 58.

[7] See *Tractatus Logico-Philosophicus*, 4.461.

[8] *An Autobiographical Study*, pp. 34–5.

[9] Ibid., p. 86

[10] Ibid., pp. 118–19.

[11] Logical positivism is itself a conventional philosophical theory that goes back to Hume: 'When we run over libraries, persuaded of these principles, what havoc must we make? If we take in our hand any volume; of divinity or school metaphysics, for instance; let us ask, *Does it contain any abstract reasoning concerning quantity or number?* No. *Does it contain any experimental reasoning concerning matter of fact and existence?* No. Commit it then to the flames: for it can contain nothing but sophistry and illusion.' *An Enquiry Concerning Human Understanding*, Section XII, Part III, p. 165.

[12] In Norman Malcolm's *Ludwig Wittgenstein. A Memoir*. p. 14.

[13] Philosophy has been epitomized in the following lines:

The Philosopher? Always defining words,
Converting the plain case into a riddle.

[14] Rush Rhees, 'Conversations on Freud', in Cyril Barrett (ed.), *Ludwig Wittgenstein, Lectures and Conversations on Aesthetics, Psychology and Religious Belief*, p. 41.

[15] It should be noted that I do not use the word 'repressed'. Wittgenstein rejected his need for psychoanalysis; it is not at all probable that he repressed it.

[16] See Charles Hanly's essay, 'Wittgenstein and Psychoanalysis', esp. p. 94, in Alice Ambrose and Morris Lazerowitz (eds.), *Ludwig Wittgenstein: Philosophy and Language.*

[17] *Philosophical Investigations*, p. 91.

[18] *Remarks on the Foundations of Mathematics*, p. 57.

[19] *A Modern Introduction to Metaphysics; Readings from Classical and Contemporary Sources.*

[20] D.D. Raphael, 'A Critical Study. *The Structure of Metaphysics* by Morris Lazerowitz', *The Philosophical Quarterly* 7, 1957, p. 80.

[21] The analogy which comes to mind is Freud's break-through to the hidden meaning of dreams. It is interesting that when Freud told his colleague Breuer that he had discovered how to interpret dreams he won as little response from him as Wittgenstein's own metaphilosophical remarks win nowadays from philosophers.

[22] Notes taken by Alice Ambrose and Margaret Masterman in the intervals between dictation of *The Blue Book.*

[23] For interesting remarks on antithetical terms see *The Blue Book*, pp. 45–6.

[24] For an extended and detailed discussion of the role of idea of feces plays in Berkeley's philosophical thought, see J.O. Wisdom's *The Unconscious Origins of Berkeley's Philosophy.*

[25] *Philosophical Investigations*, p. 47.
[26] *The Blue Book*, pp. 59–60.
[27] *On Certainty*, p. 51.
[28] Immanuel Kant, *Crtique of Pure Reason* (trans. by Norman Kemp Smith), p. 23.
[29] *Prolegomena to Any Future Metaphysics*, The Library of Liberal Arts, No. 27, p. 66.
[30] Immanuel Kant, *Critique of Pure Reason* (trans. by Norman Kemp Smith), p. 147.
[31] *Prolegomena to Any Future Metaphysics*, p. 67.

CHAPTER III

NECESSITY AND LANGUAGE

Wann man sich von der Wahrheit fürchtet (wie ich jetzt),
so ahnt man nie die volle Wahrheit. *

Wittgenstein

Underlying much of Wittgenstein's later thinking was the wish to reach a correct understanding of the nature of philosophical utterances, and this wish is also discernible in his *Tractatus*.[1] His later investigations led him to some iconoclastic ideas about what a philosophical theory is and what a philosopher does who supports his theory with an argument. Wittgenstein saw more deeply into philosophy than anyone before him had seen into it; but, for the most part, he seemed to prefer to express his perceptions in metaphorical language rather than in the language of straightforward reporting. Part of the reason for this may have been the wish to soften the hard things he saw. Remarks like 'philosophical problems arise when language goes on holiday'[2] and philosophical language is 'like an engine idling, not when it is doing work'[3] give expression to disturbing perceptions into the nature of technical philosophy but use a form of the mechanism of *sotto voce* to deflect them. Where their translations into prosaic language would tend to stir up anxiety, these words can be accepted as colorful jibes which need not be taken seriously.

It is important to notice that he stated in a number of places that philosophical propositions are not empirical. This insight into the nature of philosophical propositions (and into the modes of investigation employed in philosophy) made it fundamentally important for him to get clear on the logical difference between empirical statements and statements which have *a priori* necessity, and especially to get clear on the nature of necessity, to obtain, so to speak, an inside look into it. He appears to have come out with a conventionalist view, which on the surface at least he seems never to have given up. This is the view, generally speaking, that necessary propositions are about the literal use of terminology in a language. A number of writers have in fact described him as a conventionalist, and it must be allowed that there is considerable justification for this description. One of his frequently cited

expressions, 'rule of grammar', which he used to characterize necessary propositions, unquestionably lends some substance to the claim that he took one
of the traditional positions about logical necessitation. On one occasion G.E.
Moore, who was puzzled by the term 'rule of grammar', remarked to me that
he thought Wittgenstein meant by it what is meant by the more familiar term
'necessary proposition'. My impression at the time was that Moore thought
Wittgenstein was so using 'rule of grammar' that in his use of the term a rule
of grammar was *not* verbal. Moore's line of reasoning was perhaps the following: a rule of grammar in Wittgenstein's sense is a necessary proposition, and
since a necessary proposition says nothing about usage, a rule of grammar
says nothing about usage. There can be no doubt, however, that Wittgenstein
wished by his special use of the word 'grammar' to say that *in some way*
necessary propositions are verbal.

I

Conventionalism is open to a number of obvious objections which Wittgenstein could not have failed to know. It is worth remarking that conventionalists who are aware of these objections are not moved to give up their
position. This is mystifying and certainly calls for an explanation; for if conventionalism is taken at face value as making a factual claim about the nature
of necessary propositions, the objections are as conclusive as any objections
could possibly be. One frequently repeated criticism is that to suppose a
necessary proposition to be one which makes a declaration about verbal
usage, or one which 'records usage', is to imply that a necessary proposition is
not necessary. The negation of a true verbal proposition is a false verbal proposition, but not a proposition which could not, in principle, be true. To put
it roughly, the negation of a true verbal proposition is not a self-contradiction,
and precisely the same kind of investigation which establishes the truth of a
verbal proposition, such as recourse to dictionaries and the like, could,
theoretically, establish its denial. To use an expression of Wittgenstein's, we
know what it would be like for a verbal proposition, which happens to be
true, to be false, and for one which is false to be true. By contrast, we do not
know what it would be like for a false arithmetical *proposition* to be true, for
example, for 4 + 3 to be less than 7. Taken literally, the *philosophical* claim
that necessary propositions are about usage is refuted with complete finality

by the objection that the view that they are implies that they are not necessary.

Another well-known objection is that a necessary proposition does not say anything about terminology, because it says nothing about what language it is expressed in or about any words occurring in it. The two sentences 'Red is a color' and 'Rot ist eine Farbe' have the same meaning, which would not be the case if the proposition expressed by the English sentence made a declaration about words occurring in the sentence, and the proposition expressed by the German sentence made a declaration about words occurring in it. Wittgenstein certainly was aware of these objections, and there is reason to think that his conventionalism, which undoubtedly was the usual philosophical article at first, was transformed by his growing insight into the ways language works.

Verbal usage and logical necessity are in some way bound up with each other, and it is not too much to think that part of Wittgenstein's investigation into language and necessity was directed to bringing to the surface in what way they are bound up. Thus, in more than one place Wittgenstein remarks that a philosopher rejects a notation under the illusion that he is upsetting a proposition about things.[4] This observation shows a recognition of the difference between an explicitly verbal statement and its semantic counterpart formulated in a different idiom, an idiom which easily gives rise to the illusion that the statement is about things. The difference between a verbal proposition and a necessary proposition may only be a difference in the form of speech in which they are expressed. But the difference in the form of speech may be of great importance, and seeing this difference can lead to an understanding of the way in which language and necessity are connected. To put the matter more concretely, seeing the unlikeness, without losing sight of the likeness, between, say, the proposition that being an uncle entails being male and the proposition that being male is part of the meaning of the *word* 'uncle' can lead to a correct understanding of how the *sentence* which expresses the entailment proposition is related to the proposition it expresses.

Consider for a moment the following sentences:

(1)   A camel is a herbivore.
(2)   Ein Kamel ist ein Pflanzenfresser.
(3)   A camel is an animal.
(4)   Ein Kamel ist ein Tier.

(5)   The word 'animal' applies, as a matter of usage, to whatever 'camel' applies to

(6)   The word 'Tier' applies, as a matter of usage, to whatever 'Kamel' applies to.

The attempt to get clear on the notion of necessity requires seeing how (3) and (5) are related to each other, i.e., in what way 'A camel is an animal' is like 'The word "animal" applies, as a matter of usage, to whatever "camel" applies to', and in what way they are unlike. And to see this it is necessary to see also what (1), 'A camel is a herbivore', has in common with (3) and in what way it is different from (3). Furthermore, it is important to see how the fact that (1) and (2) have the same meaning is both like and unlike the fact that (3) and (4) have the same meaning. A clear grasp of these features of likeness and unlikeness requires seeing how (3) is related to (5), and (4) to (6). Getting a proper view of these matters will help us understand what it is about the *philosophical* view that necessary propositions are verbal, or that they state facts of usage, which makes it possible for a philosopher to hold it despite being aware of conclusive objections to it. Seeing what makes this possible will help dispel the mystery surrounding a long-standing dispute in which able philosophers with a well-preserved sense of reality can, to all appearances, debate the truth-value of a view which is known to be false without having strange ideas about each other's psychology.

Some philosophers, for one reason or another, have denied that there is a difference between logically necessary and empirical propositions, a direct consequence of which is that there is no difference in kind between the propositions expressed by 'A camel is a herbivore' and 'A camel is an animal'. Without going into the reasons for the philosophical claim that there is no difference, it can be seen that the mode of verification relevant to the proposition expressed by the first sentence is different from the mode of verification relevant to the proposition expressed by the second: observation is relevant to the first but not to the second.[5] Both sentences, equally, can be expressed as general statements of the form 'All a's are b's', i.e., as 'All camels are herbivores' and 'All camels are animals', which makes it tempting to think that both are generalizations. Their grammatical similarity appears to blind some philosophers to an important semantic difference between them. The proposition expressed by the first sentence, unlike the proposition expressed by the second, does not, to use Kant's phrase, have 'strict universality'. The first is an inductive generalization from observed instances and could in principle be upset by future instances: no number of confirming cases, however large, removes the theoretical possibility of there being a camel that is not a herbivore. By contrast, the second proposition has strict universality, which is to say that, unlike the first, it does not carry with it the theoretical possibility of being upset by a counter-instance. And this means that it is not an inductive generalization. C.I. Lewis has stated that a logically necessary proposition might, in addition to having an *a priori* demonstration, be established by 'generalization from observed instances',[6] that is, be established in the way in which a law of nature is established in science. Undoubtedly what Lewis was impressed by, and perhaps wished to highlight, is the similarity between the sentences expressing the two. But putting aside considerations of this sort, it will be clear that taken at face value his claim implies both that a logically necessary proposition of the form 'All *a*'s are *b*'s' has an associated theoretical disconfirming instance and that it does not have one. The difference between 'A camel is an animal' and 'A camel is a herbivore' may be brought into correct focus by noting that the first is re-expressible as an entailment, '*being a camel* entails *being an animal*', and the second is not, *being a camel* does *not* entail *being a herbivore*. Nothing is more plain than that being a camel is logically consistent with being a herbivore and also with not being a herbivore, and that experience alone, not penetration into the meanings of the words 'camel' and 'herbivore', will show whether it is a herbivore or not.

To come back to the philosophical claim that a necessary proposition is verbal, it can easily be seen that even though it is expressible in the form 'All neccesssary propositions are verbal', it is not put forward as a generalization which issues from an examination of instances. Instead, it is put forward as a statement to which there can in principle be no exceptions, or as one whose universality is 'strict'. Construed in this way it is restatable as an entailment: *being logically necessary* entails *being about the use of terminology*. But looked on as making an entailment-claim we are puzzled to understand the continued disagreement which revolves around it. There is no debate over whether *being a camel* entails *being an animal*; and if philosophical conventionalism did actually come down to a straightforward, elementary entailment-claim, to the effect that *being logically necessary* entails *being verbal*, there is no question but that the debate over it would have been brought to an end long ago. But if the conventionalist thesis is not to be taken as an

entailment-statement, correct or incorrect, then truly a familiar view is turned
into an enigma; we do not know *what* the conventionalist is asserting nor do
we know what we are disputing. There can, of course, be no doubt that in
some way we do understand the view and the arguments for and against it;
and the conclusion would thus seem to be that our understanding of the
view, like our understanding of our dreams, is hidden from us. No one who
lets himself become a curious observer of the philosophical scene can fail to
have the idea cross his mind that philosophy is an activity which takes place
in one of the less well lit parts of the mind. Conscious understanding of the
apparent entailment-statement should put us on the way to getting clear
about the nature of philosophical views in general. For if, as Wittgenstein has
declared, philosophical propositions are not empirical, then it is natural to
suppose them *a priori* and to be making entailment-claims. Again, as in the
case of the conventionalist position, the thing that needs to be seen is what
makes possible the continued disagreements centering on them. For example,
if the philosophical statement, 'A sense datum is private', is an entailment-
statement, it is one which is turned into a mystery by the continuing disputation over it. The only hope of dispelling the mystery and arriving at an
undistorted perception of the philosophical theory lies in getting clear on
how a logically necessary proposition is related to the sentence which
expresses it, to put it generally, how logical necessity is related to language.

In the *Tractatus* Wittgenstein makes a number of remarks about tautologies
which throw light not only on tautologies but also on all statements having
logical necessity, whether analytic or synthetic *a priori*. Proposition 6.1
states that 'The propositions of logic are tautologies', and 6.11 that 'Therefore the propositions of logic say nothing'. The view which comes out of
these two propositions is that tautologies say nothing. Usually the idea that
tautologies say nothing has been linked with the idea that they say nothing
about things, that is, with the notion that they have no 'factual content'.
Thus in 4.462 Wittgenstein states that tautologies are not 'pictures of
reality', the implication being that they give no ontological information. The
statement 'it is either raining or not raining' says nothing about the weather;
'A plant is either an oak or not an oak' says nothing about what a plant is;
'An object is either a plant or not a plant' says nothing about what an object
is. This can perhaps be made more perspicuous by considering the negations
of these statements. The expressions 'not both an oak tree and not an oak

tree', 'not both a plant and not a plant', do not function as predicates which tell us what a plant or an object is not, unlike 'not both a camel and not herbivorous', which does function to deny what a creature is. To say with regard to anything that it is not both a camel and not a herbivore is to say what the thing is not, and this is because the predicate 'both a camel and not a herbivore' presents a possible 'picture' of the thing. But to say with regard to a plant that it is not both an oak tree and not an oak tree is not to say what the plant is not, inasmuch as 'both an oak and not an oak' does not have a use to describe any plant, actual or hypothetical. Tautologies say nothing about what there is and what things are like, and contradictions say nothing about what there is not and what things are not like. Predicates of the form 'ϕ or not ϕ' equally with 'ϕ and not ϕ' have no descriptive content.

These considerations apply to all analytic propositions, and to synthetic *a priori* propositions as well. Kant and many philosophers after him have held that synthetic *a priori* propositions, the predicates of which are connected by 'inner necessity' to their subjects but are not components of them, have factual content, that is, delineate features of the world. But it will be clear that a true proposition, and hence a logically necessary proposition of whatever kind, will tell us something about what there is only if its negation states something to be the case which in fact is not the case. Kant's claim that the proposition that every change has a cause is *a priori*, although not analytic, has, as is known, been challenged and debated over and over again with a vigor which promises the debate immortality. Without going into its detail, we may say that if *being a change* entails (whether synthetically or otherwise) *having a cause*, then *being a change and not having a cause* will not be a predicate of any conceivable occurrence and thus will not have a function to describe what does not take place. The conclusion would seem to be that the proposition that a change must have a cause, like a tautology, says nothing about what does or does not take place. Consider for a moment the proposition 'A red thing is not green'. It clearly is an *a priori* truth and it could be argued, in the following way, that it is also synthetic. *Being red* entails *not being green*, but the concept *not green* is not discovered by a 'dissection' of the concept *red*: the consequent concept is not a conjunctive part of the antecedent concept. In other words, it could be argued that the consequent is synthetically entailed by the antecedent, and thus that the proposition, 'A red thing is not green', is synthetic *a priori*. But as in the case of a tautology, it says nothing about what things are, if its negation does not present a picture of a hypothetical reality. And since *being red and simultaneously green* is not a predicate of a conceivable object, *not being both red and green* will fail to function as a predicate which says what a thing is not.

It becomes clear now that Wittgenstein's claim that tautologies say nothing can be extended to all propositions which have *a priori* necessity. With regard to tautologies he said they are 'not, however, nonsensical. They are part of the symbolism, just as "0" is part of the symbolism of arithmetic' (4.4611). The implication of this would seem to be that a sentence which expresses a proposition that says nothing is not a nonsensical combination of words like 'Duplicity reclines on the first odd prime'. It would generally be maintained that an indicative sentence which is not nonsensical, that is, one which is literally intelligible, must say something. To put it equivalently, a sentence which says nothing whatever would be said to be nonsensical or to have no literal meaning. The idea behind this claim is that an intelligible declarative sentence must be about something, actual or imaginable, that it must have a subject about which it makes a declaration. A sentence which expresses an empirical proposition puts forward a claim about the world; it says something about what in fact is the case or is not the case and has some sort of subject of discourse. A tautology, which says nothing, but nevertheless is not non-sensical, must therefore have some subject matter about which it makes a declaration.

Some philosophers have identified the subject of *a priori* statements as the structure of the world. In his *Notebooks* of 1914–16 Wittgenstein wrote: 'The great problem about which everything I write turns is: Is there an order in the world a priori, and if so what does it consist in?'[7] And in the *Tractatus* there is the suggestion that the subject of tautologies is identified as the structure of the world. In his own words, 'The propositions of logic describe the scaffolding of the world, or rather they exhibit it. They have no "subject matter."... It is clear that something about the world must be indicated by the fact that certain combinations or symbols – whose essence involves the possession of a determinate character – are tautologies' (6.124). The implication of these words is not that tautologies have no subject matter but rather that their subject matter is not of a certain sort. They are 'about the world' in a particular respect, namely, about its basic structure, and this, not the specific contents of the world, is their subject matter.

This notion would seem to be in accord with Leibniz' view that necessary propositions are true for all possible worlds. Leibniz' distinction between necessary, or identical truths, and truths of fact is that the latter hold only for particular possible worlds, true for some and false for others, while identical truths hold for every world, for the existing world as well as for non-existing possible worlds. The underlying idea is that an *a priori* truth has some sort of ontological import. It is about the world, just as an empirical proposition is, but it is not only about this world. We may gather that Leibniz had the idea, whether or not he was fully aware of it, that an *a priori* necessity refers to that which is invariant in all possible worlds, to what might be called a cosmic constant, and this is the structure to which all possible worlds conform.

Wittgenstein's way of putting this is summed up in 6.12: 'The fact that the propositions of logic are tautologies shows the formal — logical — properties of language and the world.' Some philosophers have characterized the three Aristotelian laws of thought as laws not only to which thinking must conform but also as laws to which things must conform. The proposition that nothing can be both an oak tree and also not an oak tree is a different kind of law from a law of physics, e.g., the law that every particle of matter is attracted to every other particle with a force that varies directly as the product of their masses and inversely as the square of their distance apart. A law of logic may be said to apply to special laws of nature as well as to the specific characteristics of things. The idea behind the views of Wittgenstein and Leibniz is that any system of things together with the laws governing their behavior, however different from any other system of things and their laws, will fall under the same laws of logic, and more generally, the principles embedded in all *a priori* propositions. The cosmic picture linked with this idea is that *a priori* statements give the structure to which all things and laws, actual and possible, must conform. Contingent truths refer to the contents of the present cosmos; the totality of *a priori* truths details its logical structure. Thus, equally with empirical propositions, tautologies have a subject: the invariant structure of all possible worlds. Wittgenstein's two statements, 'Tautologies say nothing' and 'They are not nonsensical', would seem to imply on his own accounting that tautologies say nothing about what there is in the world but nevertheless do have a subject: the structure of reality which they explicate, or in some way reveal.

A philosopher who perceives that a tautology says nothing, for example, that 'It is either raining or not raining' says nothing about the weather, but who does not deny that it is intelligible will, if his thinking is governed by the formula 'An intelligible statement cannot be about nothing', *find* something which it is about. For a time Wittgenstein identified the subject of *a priori* truths as the structure of the world, but this seems to have been only a transitional view. The insight that necessary propositions have no factual content may develop into the perception that they convey no sort of information whatever about the world; and this seems to have occurred in Wittgenstein's thinking. One consideration which shows that tautologies have no factual content also shows that they are not about the structure of the world either. The negation of a logically necessary truth presents us, in John Locke's words, with an 'impossibility of thought'. The negations of 'A red thing is not also green' and of 'A physical particle which is in one place is not at the same time in another place' result in combinations of terms which stand for impossibilities of thought, namely, the expressions, 'a red thing which is green' and 'a physical thing which is in two different places simultaneously'. So to speak, they present us with conceptual blanks. It will be clear that if these phrases denoted concepts instances of which we could imagine, then instead of denoting impossibilities of thought they would present us with conceivabilities, that is, with 'thinkable states of affairs'.[8]

To suppose, however, that they apply to hypothetical instances is to imply that it is possible to conceive of what would upset a necessary proposition. It would thus imply that a necessary proposition is in principle falsifiable. It would imply that we know what it would be like for there to be a red thing which is also green, for a plant to be both clearly an oak tree and also definitely not an oak tree, and for one and the same oak tree to be in a given place and also elsewhere, and hence that we can envision circumstances which would make true a self-contradictory proposition and make false a logically unfalsifiable proposition. Wittgenstein has remarked (3.031): 'It used to be said that God could create anything except what would be contrary to the laws of logic. − The reason being that we could not *say* what an "illogical" world would look like.' To this we might add that to deny that God could create something contrary to the laws of logic, or a self-contradictory state of affairs, is not to imply that there is something which God cannot do. For a putative descriptive expression which involves a contradiction has no descriptive content, i.e., has no use to describe anything, actual or not. Hence, to say that God can create nothing which goes against the laws of logic, or that God cannot create a state of affairs which answers to a self-contradictory expression, is not to use language to state what cannot be done. Wittgenstein appears to have intended this in 3.032 when he says 'it is. . .impossible to represent in language anything that "contradicts logic" '.

The implied difference between a phrase which denotes a logical impossibility and an expression which denotes a physically impossible state of affairs, one which, if it existed, would cancel an immutability in nature, is that the second presents us with a thinkable state of affairs, to use Wittgenstein's word, a state of affairs we can *picture* to ourselves (3.001), and the first does not. Thus the negation of a necessary proposition neither shows nor represents nor exhibits nor depicts what the structure of the world cannot be. Hence a necessary proposition does not represent or exhibit or depict a structure that the world must have. If the one does not show *what* the structure of reality cannot be, or perhaps better, does not exhibit a structure to which reality could not conform, then the other does not depict a structure to which reality must conform. The statement, 'A thing can be in one place only at a given time', would tell us something about what must be with regard to things in space only if 'A given thing is in two separate places simultaneously' *described* what cannot occur in space. In this regard 'A thing must be in one place only at a given time' is completely different from 'A thing is gravitationally influenced by other things in space'. The second describes what happens in space, the first does not.

Giving up the idea that the subject matter of tautologies is the structure of the world does not mean giving up the idea that they must have some sort of subject matter. Parenthetically, it is not difficult to see that the expression 'the logical structure of the world' is a made-up expression to which no clear meaning has been assigned. Its apparent function is to serve as the 'name' of the subject matter of tautologies, devised in the course of looking for the subject matter under what might be called the 'regulative' formula that a literally meaningful statement must be about something. A philosopher who thinks that tautologies are in no way whatever about reality and who rejects the metaphysical claim that the meanings of general words are abstract entities might then fix on the use of terminology in a language as the subject matter of tautologies, and in general of *a priori* statements.

A philosopher like Wittgenstein, who later saw more deeply into the workings of philosophy than anyone else and eventually arrived at the idea that a philosopher rejects a notation under the delusive impression that he is up-setting a proposition about things, will in the course of his intellectual odyssey try out various 'theories' regarding the subject matter of *a priori* statements. Wittgenstein certainly tried out conventionalism, which represents the use of terminology as what necessary statements are about.

Elsewhere I have tried to show that a philosophical theory is a gerry-mandered piece of terminology which, because it is presented in the onto-logical form of speech, tends to create the illusion that conceals what is being done with language. Without going into this here, it needs to be pointed out that an ontologically presented re-editing of terminology can have either of two purposes, which do not necessarily conflict with each other. It can have the purpose of highlighting in a graphic way a point of usage; it can also have the purpose, undoubtedly unconscious, of creating an illusion that a theory about things is being advanced. In Wittgenstein these two purposes do not stand out in clear separation from each other. It is safe to say that the conventionalist position which he sometimes took is the usual variety, which itself is a philosophical theory, that is, a 'theory' with a built-in possibility of endless disputation. The insight which goes beyond conventionalism, and does not issue in a philosophical theory, is to be found in his later writings; but he never presented it in clear articulation, unshadowed by metaphor.

Traditional conventionalism is one of the theories which appears to be adopted in the *Tractatus*. Thus, he wrote (6.126): 'One can calculate whether a proposition belongs to logic by calculating the logical properties of the *symbol*. And this is what we do when we "prove" a logical proposition. For, without bothering about sense or meaning, we construct the logical propo-sition out of others using only *rules that deal with signs*.'[9] This position also makes its appearance in some of his later work. Moore, in discussing some things Wittgenstein said about tautologies in his lectures 1930–33, suggests that Wittgenstein identified a statement of the form 'it is logicallly impossible that p' with the statement 'the sentence "p" has no sense'.[10] Moore went on to remark, 'why he thought (if he did) that "It is logicallly impossible that p" means the same as "The sentence "p" has no sense", I cannot explain'.[11] He also reports him as having stated that 'the proposition "red is a primary colour" was a proposition about the word "red"; and, if he had seriously held this, he might have held similarly that the proposition or rule "$3 + 3 = 6$" was

merely a proposition or rule about the particular expressions "3 + 3" and "6" .'
Moore observed that 'he cannot have held seriously either of these two views,
because the *same* proposition which is expressed by the words "red is a primary colour" can be expressed in French or German by words which say
nothing about the English word "red"; and similarly the *same* proposition or
rule which is expressed by "3 + 3 = 6" was undoubtedly expressed in Attic
Greek and in Latin by words which say nothing about the Arabic numerals
"3" and "6". And this was a fact which he seemed to be admitting in the
passage at the end of (I).' [Notes of lectures in the Lent and May terms of
1930] .[12]

It is certainly not a rare thing for a philosopher to hold a view while aware
of 'fatal' objections to it. This blitheness of attitude toward refuting evidence
is not encountered in the sciences, and its occurrence in philosophy stands in
need of explanation. In this connection, Moore's paradox forces itself on our
attention: 'The strange thing is that philosophers have been able to hold
sincerely, as part of their philosophical creed, propositions inconsistent with
what they themselves *knew* to be true. .'[13] It may be unkind, but it appears
to be true, to say that philosophers have not sincerely faced up to this paradox. Instead of being made curious about the nature of their activity, about
what it is they are doing with words, they push the paradox out of their mind
and go on doing philosophy with what seems to be a determined lack of
curiosity. Be this as it may, there is no question but that Wittgenstein at times
adopted a conventionalist view about *a priori* necessity, though he did not
remain irremovably attached to it. Conventionalism does represent insight
into the nature of logically necessary propositions, but presented in the form
of a *theory* about *a priori* necessitation it is an obstacle to the understanding
of the nature of philosophical statements. Wittgenstein overcame this obstacle
and arrived at an understanding of how philosophy works. The following report of the way he began to think about language and necessity shows clearly
his growing perception into the special way a sentence which expresses a
necessary proposition and the necessary proposition are bound up with each
other:

. . . .'4' has two different uses: in '2 + 2 = 4' and in 'there are 4 men here'. . .We must
understand the relation between a mathematical proposition about 4 and an ordinary
one. The relation is that if the word '4' is a word in our language, then the mathematical
proposition is a rule about the usage of the word '4'. The relation is between a rule of

grammar and a sentence in which the word can be used. . . . Now is '2 + 2 = 4' about 4 or not? If the sentence 'I have 4 apples' is about 4, then '2 + 2 = 4' is not about 4 in this sense. If you say it's about the *mark* '4', be careful. . .When do I emphasize the word 'rule'? When I wish to distinguish between '2 + 2 = 4' and 'There are 4 apples on the table'. . .If I use the word 'rule' it is because I wish to oppose it to something else. . . If 'There are 2 men here' is about 2, then to say '2 + 2 = 4' is about 2 is misleading, for it's 'about' in a different sense. . . . . .25 × 24 = 600 isn't used as a rule for handling signs, though it would stand in the relation of a rule to a *proposition* using this equation.[14]

The objections to conventionalism, construed as making a claim regarding what necessary propositions are about, that is, regarding what their 'subject matter' is, are conclusive. It is unrealistic to think that the theory, taken at face value, could be held by anyone who was aware of them: it would require our having to think that the conventionalist was suffering from an odd mental condition which enabled him in some way to seal off the objections to the view from the view. Looked at through the spectacles of Moore's Paradox, we should have to imagine that a philosopher believes a view to be true while aware of objections which he knows show it to be false. The conclusion forced upon us, however strong our resistance to it may be, is that despite its appearance of making a claim regarding what necessary propositions are about, it makes no such claim. To understand what the conventionalist is trying to bring to our attention, consider three of the six sentences given earlier:

(1)   A camel is a herbivore.

(3)   A camel is an animal.

(5)   The word 'animal' applies, as a matter of usage, to whatever 'camel' applies to.

The difference between (1) and (5) is obvious. (5) is about the word 'camel' and (1) is about what the word 'camel' denotes. That is, to put it roughly, one is about a word and the other about a thing. (3) is neither about the word 'camel' nor about what in (1) is denoted by the word 'camel': it is neither about a word nor a thing. What we know in knowing that what (1) says is true is something about camels. What we know in knowing that what (5) says is true is a fact of usage, and what is known in these two cases is something in addition to our understanding the sentences. In this regard (3) is different from both (1) and (5). Understanding it is equivalent to knowing a fact about verbal usage, although this fact is not expressed by the sentence. (3) shares its form of speech with (1), the ontological idiom in which words are not mentioned and are usually used to refer to things. Its content, however, what might be called its invisible subject matter, is shared with (5).

Perhaps this point is best brought out in the following way: The fact that the sentence 'A camel is an animal' expresses a necessary proposition is equivalent to the fact that the sentence 'The word "animal" applies, as a matter of usage, to whatever "camel" applies to' expresses a true verbal proposition. To put it somewhat metaphorically, the verbal content of (5) is explicit and visible, while the verbal content of (3) is hidden; it is made invisible by the mode of speech in which the sentence is formulated. (1) has factual content, (5) has verbal content, and (3) has hidden verbal content. (3) is a grammatical hybrid which is sired by (1) and (5) and differs markedly from both. As an aside, it may be observed that philosophical theories about the nature of necessity are nothing more than academic assimilations of the offspring to one or other of the parents, with the consequence that its relation to one parent is, to use Wittgenstein's expression, hushed up.

The objections to conventionalism can now be seen to call attention to respects in which (3) is different from (5), and there is a temptation to take them to be objections against identifying (3) with (5). But if we can resist retreating into philosophical fantasy, we can understand how the conventionalist is able to hold his position only by supposing that he has made no such identification, however much his words suggest that he has. Supposing that the conventionalist does not in fact make this identification requires our thinking that conventionalism is not a description of the subject matter of necessary statements. Instead, it is to be construed as a way of highlighting the likeness between (3) and (5), in disregard both of their difference and of the likeness between (3) and (1). With his pronouncement that necessary propositions are really verbal the philosopher heightens a similarity which seems unimportant to him. This he does by artificially stretching the use of the word 'verbal' so as to cover, if only nominally, necessary propositions. We might say, for the purpose of bringing out the point, that the word 'verbal' has two uses which are made to appear as the same use: the original use to describe the nature of some propositions, and a new descriptively empty use with what might be called a grammatical point. By means of this stretched use he brings nearer to us a similarity, while keeping at a distance a dissimilarity. Being a philosopher he dramatizes what he does by presenting it in the guise of a theory, to which, it must be said in his defence, he himself falls dupe. Instead of saying 'a camel is an animal' is like 'the word "animal" applies to whatever "camel" applies to' but is unlike it in not mentioning words, and that it is unlike 'a camel is a herbivore' in not being about camels, he says 'The proposition *a camel is an animal* is *really* verbal'. When a philosopher uses the word 'really' he appears to be reporting a discovery, whereas, as Wittgenstein remarked, 'what he wants is a new notation'.[15] A new use of 'verbal' is presented in a way which creates the impression that the true nature of necessary propositions is being revealed.

To return for a moment to Moore's objection against saying that the proposition that $3 + 3 = 6$ is about the expressions '$3 + 3$' and '6', the objection, namely, that the same proposition which is expressed by '$3 + 3 = 6$' is expressible in other languages by words which say nothing about the Arabic numerals '3' and '6'. It is a fact that the sentences '$3 + 3 = 6$' and 'drei und drei macht sechs' mean the same, or express the same proposition; and they would not if the first said something about the use of '3' and '6' and the second said something about the use of 'drei' and 'sechs'. For the first says nothing about the German words and the second says nothing about the Arabic numerals. Since the two sentences express the same proposition, neither sentence can say anything about the symbols which occur in it. As is known, some metaphysically inclined logicians have adopted a view according to which these sentences and others like them are about abstract entities.[16] To revert to sentences (1)–(6), one difference between 'A camel is a herbivore' and 'A camel is an animal' is, on the Platonic theory, that the first is about camels and the second about abstract camelhood. Here no more can be said about the metaphysical difference of 'subject matter', in the one case things, and in the other case supersensible entities, than that the Platonic theory, like the cosmic structure theory, is the product of thinking which is governed by the formula that an intelligible indicative sentence must be about something.

To continue the explication of Moore's point by reference to the sentences (1)–(6), it is correct English to say that (1) and (2), that is, 'A camel is a herbivore' and 'Ein Kamel ist ein Pflanzenfresser', have the same meaning. It is correct English to say that (3), 'A camel is an animal', and (4), 'Ein Kamel ist ein Tier', mean the same; but it is not a correct use of English to apply the phrase 'mean the same' to the pair of sentences, (5), 'The word "animal" applies, as a matter of usage, to whatever "camel" applies to', and (6), 'The word "Tier" applies, as a matter of usage, to whatever "Kamel" applies to'. Sentences (1) and (2) have a subject matter, as also do sentences (5) and (6); but (3) and (4) have only a contrived subject matter, which is to say they are made to appear to have a subject matter. Neither (3) nor (4) says anything either about words or about things, and this fact has led some philosophers to hold that they say nothing. Sentences (1) and (2) say the same thing about the same subject. This feature of the pair of sentences gives us one condition for the correct application of the phrase 'mean the same'. Sentences (5) and (6) do not satisfy this condition: they say similar things about their subjects, but their subjects are different, which makes it incorrect to apply the phrase 'mean the same' to them. By contrast, (3) and (4) have no subjects; but it is correct English, nevertheless, to say that they mean the same. They have the same meaning, although they make no declaration about anything. We might, in order to bring out a point, say that 'mean the same' does not have the same meaning in its application to (1) and (2) that it has in its application to (3) and (4). Although (1) and (2) translate into each other, and (3) and (4) like-wise, (1) and (2) are about the same subject, while (3) and (4) have no subject.

Nevertheless, it is not true that (3) and (4) say nothing or that they are literally meaningless. To put the matter briefly, understanding them comes down to knowing facts about the use of terminology, although terminology is not the subject of these assertions. The sentence 'A camel is an animal' is a grammatical crossbreed with one foot, so to speak, in the correlated verbal sentence in the same language and the other in related nonverbal fact-claiming sentences. (3) and (4) translate into each other, which is a feature that makes it correct to apply the term 'mean the same', or 'have the same meaning', to the sentences, despite their not being about anything.[17] But a person who understands both sentences will know facts of usage in different languages, while a person who understands only one of these sentences would not know the fact of usage exhibited by the other. The phrase 'mean the same' is used to refer to one feature when applied to sentences in the hybrid idiom and is used to refer to a further feature when applied to fact-claiming sentences. But by artificially equating 'mean the same' with 'say the same thing about the same subject' – under the rule that a literally meaningful indicative sentence *must* be about something – a philosopher creates the illusion that (3) and (4) have a special rarified subject matter, that they are about supersensible objects which can only be grasped by the pure intellect. A comment of Wittgenstein's is worth noting in this connection: ' "The symbol 'a' stands for an ideal object" is evidently supposed to assert something about the meaning, and so about the use, of "a". And it means of course that this use is in a certain respect similar to that for a sign that has an object, and that it does not stand for any object.'[18] When a philosopher like Moore points out that (3) and (4) mean the same and therefore that neither could mean the same as (5) or (6), he is calling attention to a grammatical similarity between sentences which express necessary propositions and those which make factual claims about the world, while pushing into the background the likeness between sentences which express necessary propositions and those which express propositions about usage.

Wittgenstein sometimes characterized necessary propositions as rules of grammar, with the idea in mind that he was stating a theory about the nature of logical necessity. He was able to see past this idea, however, and at other times he called them 'rules of grammar' in order to direct our attention to an important feature of sentences which express necessary propositions. Probably part of his reason for wishing to accent the verbal aspect of necessary propositions was to remove the idea that they are about mysterious things and thus to dispel the occult air which tends to settle over them. But more important than this, he undoubtedly felt that getting a clear understanding of the nature of necessary statements is required for getting clear about how philosophy works. Seeing what breed of theory conventionalism is, which comes to seeing what a conventionalist does with the word 'verbal', or equivalent expressions, is the final step towards understanding the theories of philosophy. The conventionalist theory is one of a large family of theories, and to get an inside look into it is to get an inside look into the other members of the family.

Wittgenstein has said that what the philosopher needs in order to solve, or to 'dissolve', his problems is to 'command a clear view of our use of words'.[19] We might add that what he needs is an improved understanding of sentences which express or are put forward as expressing necessary propositions, and to see how they both conceal their verbal content and also create the impression of being about phenomena. It will be recalled that Wittgenstein stated that philosophical theories are not empirical, and that he also said that a philosopher rejects an expression under the delusive impression that he is upsetting a proposition about things. This idea about the nature of philosophical theories and philosophical refutations is deeply rooted and according to him 'pervades all philosophy'.[20] The conventionalist theory will be recognized now as falling under Wittgenstein's characterization. It is not empirical; it is restatable as making the entailment-claim that being a necessary proposition entails stating verbal facts. Reformulating it as an entailment-statement dissolves the notion that it is an inductive generaliz- ation, and it also changes our idea of what a philosopher is doing who, to all appearances, is demolishing by unanswerable arguments a widely held theory about what a necessary proposition is. It brings into clearer view what the theory is not and it also puts us on the way to a correct understanding of what it is.

It is easy now to see why the conventionalist can hold his position against all conclusive objections. He is presenting a stretched use of 'verbal', a use which artificially covers necessary propositions, and is not using the word 'verbal' in the normal way to make a false statement about them. Equally with the philosopher who satisfies himself that he has refuted the conven- tionalist theory, the conventionalist can survive refutation after refutation and remain satisfied that, all the same, his theory is not incorrect. Both the philosopher who holds the conventionalist theory and the philosopher who rejects it suffer from the fallacy which pervades all philosophy: namely, the false notion that the dispute centers on the truth-value of a theory rather than on the academic redistricting of a term.

II

This understanding of what the conventionalist theory comes to has direct application to the philosophical problem of the privacy of experience. Con- sider the following words: 'When philosophers assert that experiences are private, they are referring to a necessary proposition. It would be a contra- diction to speak of the feelings of two different people as being numerically identical: it is logically impossible that one person should literallly feel another's pain. But these points of logic are based on linguistic usages which have, as it were, the empirical facts in view. If the facts were different, the usage might be changed.'[21] When we consider these words with care, we discern in them three claims. One, explicitly stated, is that the sentence 'experiences are private' expresses a necessary proposition. Another is the conventionalist view that the necessary proposition is really verbal, i.e., 'the points of logic' are 'based on linguistic usages'. And still a further claim is that in some way a matter of fact about experience is involved. What comes through quite distinctly in the words that the 'points of logic are based on linguistic usages which have, as it were, the empirical facts in view', and that 'if the facts were different, the usage might be changed', is the idea that experience is private, as a matter of empirical fact.

It is by no means uncommon for a philosopher to hold explicitly that his utterance expresses an *a priori* truth and also to imply that it refers to empirical fact; and in the present case the implication stands out in bold relief. To put the matter shortly, what is being held is that the proposition expressed by 'Experience is private' both is logically necessary and also has 'factual content', i.e., makes a factual claim about feelings, pains, and the like. Wittgenstein's observation that philosophical problems are not empirical carries with it the implication that philosophical answers to philosophical questions are not empirical. But it should no longer be necessary to remark that the philosophical statement that experience is private has its absorbing and continued interest for philosophers because, for one thing, of the empirical picture associated with it, the picture of our having experiences which no one is privileged to share with us.

The contradiction implied by the conjunction of philosophical claims is blatant, and it parallels the contradiction frequently pointed out in the view that necessary propositions are really verbal, or that they are 'based on linguistic usages'. Philosophers are not overly troubled by contradictions like these; indeed, such contradictions become permanent additions to the content of philosophy. Anyone who is realistic about philosophy will find it hard not to think of it as a growing collection of contradictory theories which are not given up and of paradoxes which remain in permanent suspension. A further contradiction can now be added to the collection. This is the contradiction that is implied by the conventionalist view of logical necessity in combination with the tacit claim that at least some necessary propositions are about things, in the present instance, the contradiction which comes out of holding that the proposition that experience is private is 'based' on the use of terminology and that it also states a matter of fact. To put the matter in terms of the *sentence* 'Experience is private', holding that it expresses a necessary proposition amounts, on the conventionalist thesis, to stating that it is *about the use* of the terms 'experience' and 'private', and thus that the sentence *does not use the terms* to make a statement about what they are ordinarily used to refer to. The contradiction which emerges is that a sentence which is about words rather than things is nevertheless about things. Perhaps a more perspicuous way of making this contradiction explicit is the following. On the conventionalist view, to hold that the sentence 'Experience is private' expresses a necessary proposition comes to holding, in part at least, that the phrase 'non-private experience' has been given no application to anything, i.e., that it has no descriptive function in the language. And to hold, in immediate conjunction with this, that the sentence refers to an empirical fact about experience, is to embrace the contradiction that a phrase which has no descriptive use nevertheless has one. It is important to try to understand what makes it possible for a philosopher to accept this and related contradictions, and treat them as if they somehow do not go against their views. The insight reached into the nature of the conventionalist view helps us toward an understanding of why a contradiction in mathematics eliminates a proposition but does not do this in philosophy.

The passage cited from Ayer above tacitly implies that the phrase 'feels another's pain' refers to what is logically impossible, and thus that it does not have a use to describe a conceivable occurrence. It states, also, that if the empirical facts were different, usage might be changed, which is to say that the phrase might then be given a descriptive use. It is not clear what the empirical facts are which the linguistic usages keep in view. The only facts which, so to speak, *fit* the case are those which would be described by 'feels another's pain', and which, if they obtained, might make us *give* a descriptive use to this combination of words. The passage is labyrinthine in its ambiguity, but what makes itself evident is the idea, which probably all philosphers have, that an expression for a logical impossibility describes what never in fact happens. One philosophical logician has said: 'That which necessarily is the case is also as a matter of fact the case',[22] and it is fair to infer that he also has the idea that what is logically impossible never occurs as a matter of fact. It cannot be pointed out too often that a phrase which expresses what is logically impossible, e.g., the phrase 'soundless crash of thunder', does not have a use to *describe* what is not or what cannot be: it simply has no descriptive function in the language in which it occurs. It may be useful to point out that the proposition that an expression which denotes a logical impossssibility has no descriptive content is not itself an experiential proposition, but rather declares an entailment, to the effect that being an expression for a logical impossibility entails being devoid of descriptive content. The assertion that an expression refers to what is logically impossible is incompatible with the assertion that it describes what in fact never occurs: if it describes what does not occur, it does not refer to a logical impossibility, and if it refers to a logical impossibility, it does not describe what does not happen.

If the sentence 'Experiences are private' does refer to a necessary truth, in virtue of the ordinary conventions governing the use of 'private' as well as of experience-denoting terminology,[23] then it makes no declaration about experiences. It exhibits, without expressing, what is stated by the sentence 'The word "private" correctly applies to whatever "experience" applies to, such that the phrase "an experience which is not private" has no descriptive use'. And if the phrase 'feels another's pain'· does, in virtue of the rules governing the words occurring in it, refer to what is logically impossible, then it does not describe what one cannot feel, or feelings that one cannot have. Now, a philosopher who holds both that 'feels another's pain' has no descriptive use and also that it describes what never happens, and declares that if what does not happen were to happen we might then *give* a descriptive sense to 'feels another's pain', would seem to have lost his way. It is tempting to think that anyone who states that a term which lacks a descriptive use might be given one if situations answering to it came into existence is making a mere mistake, which for some odd reason he fails to see. The terms 'eternity bone' and 'phlogiston' fell into disuse when it was finally decided that nothing existed which answered to them, and, undoubtedly, if the facts had been different the terms would not have fallen into disuse. But these cases are not comparable to the philosophical case, which is like that of being told that the expression 'prime number between 13 and 17' might be given an application if such a number were discovered.

There is a strong temptation to think that our opponents in philosophy make mistakes which are plainly visible to us but which they do not have the wit to see. Hardly any philosopher fails to succumb to it, not only because it makes him feel superior but also because it sustains the truth-value facade which hides from him the real nature of philosophy. The importance for the

correct understanding of philosophy of putting aside the truth-value spectacles through which philosophers look at their work will be realized by
anyone who is not on the defensive about the unexplained difference
between the chronic condition of philosophy and the condition of the experimental and mathematical sciences. To put the matter in terms of a probability
evaluation, it has now become more probable that philosophical assertions
have no truth-value than that they do have truth-values which philosophers
have been unable to agree on.

If we keep separate the statement that it is logically impossible for one
person to feel another's pain from the statement that no one as a matter of
fact ever feels another's pain, and suspend the idea that a mistake is involved,
we can see how the two statements work with respect to each other to
produce the philosophical theory that experience is private. Consider first the
claim that it is logically impossible to feel another's pain, which is linked
with and in fact derives its importance from the empirical picture of contents
which are not accessible to more than one person, comparable in some ways
to the contents of a bank box to which only one person has the key. If the
claim about the impossibility of feeling another's pain is an *a priori* truth,
understanding a sentence which expresses it is equivalent to knowing that
the descriptive part of the sentence (the phrase 'feels another's pain', in the
English sentence) has been assigned no use to describe anything.

It will be clear thus that if the phrase 'feels another's pain' expresses what
is logically impossible, the words 'another's pain' do not have a use in the
language to distinguish between pains that a person is able to feel and pains
he is prevented, for one reason or another, from feeling. It will also be clear
that a philosopher who asserts the imposssibility of feeling another's pain
draws from his assertion the consequence that a person can feel only his own
pains. The background picture linked with the consequence is that of someone who is *confined* to feeling certain pains. But the picture turns out to be
inappropriate to the words 'A person can feel only his own pains'. For 'feels
his own pain' is semantically connected with 'feels another's pain' in such a
way that if the second expression has no use to describe anything, neither
does the first. The terms 'another's pain' and 'his own pain' serve to make a
contrast, such that if in a certain context 'another's pain' describes nothing,
in the same context 'his own pain' will describe nothing. Thus, if in the
sentence 'By contrast to another's pain, which I cannot feel, I can feel my

own pain', the phrase 'another's pain' does not describe what I cannot feel, then the phrase 'my own pain' does not describe what I can feel. Put somewhat differently, 'his own pains' will have a use to set off pains a person can feel from pains he does not or cannot feel only if 'another's pains' also has a use to set off pains a person does not or is unable to feel from those he does or is able to feel. To imitate Bishop Butler, the sentence 'A person can feel his own pain and not another's pain' says nothing about what a person can feel, if the required contrast is cancelled by the failure of one of the terms to have a descriptive function. Either both terms have a use or neither term has a use. Hence if the philosopher of the privacy of experience followed through on the consequences of his claim with regard to 'feels another's pain', he would have to allow that 'feels his own pain' has no use to describe what a person is limited to feeling. But this, quite obviously, he does not wish to do. If he did maintain, whether explicitly or indirectly, that both expressions lacked descriptive sense, his 'theory' would vanish, cancel itself out of existence.

Wittgenstein has characterized as a typically metaphysical mistake[24] the use of one of a pair of antithetical terms in what might be called linguistic dissociation from its antithesis, i.e., retaining one of the terms while deleting the other by the artful technique of 'showing' by an argument that the other stands for a logical impossibility. He described at least some of his later work as consisting of 'bringing words back from their metaphysical to their everyday usage'.[25] The semantic fact about antithetical terms is that if one of a given pair is stripped of its use in the language, without being replaced by a term to do its work, the other also loses its use. By bringing back words from their metaphysical to their everyday use, which is to say, by restoring to them their former use, the words which normally function as their antitheses *recover* their use. Wittgenstein's language sometimes suggests the idea that the metaphysical use of a term, as against its everyday use, is an actual use which assigns to a word the role of describing occult realities or real as against merely apparent states of things. It is clear, however, that a metaphysical use is not given to a term *independently* of the semantic suppression of its antithesis. Instead, its metaphysical use is *the result of*, and thus is created by, the ontologically reported suppression of the word. Instead of speaking of the metaphysical use of a term it would be less misleading to speak of the metaphysical job a term acquires when its antithesis is (academically) cast out of the language. When one of a pair of antithetical terms is suppressed the other loses its function to describe and takes on metaphysical, illusion-creating work.

A term which is shorn of its descriptive use by making its antithesis stand for a logical impossibility does not spontaneously acquire a new descriptive use. Any new use it has it must be given, and if it is not in fact given a use, the use it appears to acquire is one in appearance only. The problem is to explain how the illusion is brought about of its having a use to describe reality, how, in the present connection, the appearance is created that a dramatic claim is being made by the words 'A person can feel only his own pains'. If we go back to our interpretation of the conventionalist position, we can get a behind-the-scenes look at the semantic props which are used to bring to life the scene at the front of the philosophical stage. The props which produce the delusive picture of everyone being enclosed by a wall over which no one else can look are not either the wrong use of 'feels another's pain' and 'feels only his own pain', nor the mistaken descriptions of their actual use in the language. The props are academic, linguistically idle alterations, which when held up against everyday, unrevised language, give rise to the delusive impression that a fact about the nature of feelings and sensations is being disclosed.

The assertion that it is logically impossible to feel another's pain, which is linked with the declaration that a person is limited, by logical necessity, to feeling his own pain, embodies a piece of gerrymandered terminology. The result is to *deprive* an expression of its use, as part of a game that is being played with language. The fact that the game is conducted in the nonverbal mode of speech conceals its verbal nature. What is being done with words is hidden all the more effectively by the fact that the use they have in everyday language *remains intact* in everyday language, which thus serves as a backdrop and gives to the game the appearance of a discovery. Wittgenstein has said that a philosophical problem arises when language goes on holiday, and it seems that the philosophical view that experience is private is, as we might say, an image thrown on the language-screen by an ontologically presented, non-workaday revision of grammar.

It is unrealistic to think that the verbal game by itself is so entirely captivating as to make it worth the candle. Its ability to keep thinkers in permanent intellectual thralldom requires us to suppose that something else is involved, something which it is difficult to recognize consciously. A brief speculation is permissible at this point, and it may be enlightening. The suggestion that part of the mind is a dark area whose contents are inaccessible to us tends to be received not only with the expected rejection but also with a kind of grudging fascination. Philosophers perhaps more than other intellectuals resist the idea that the mind contains a submerged Atlantis; but there can be no doubt that they too, and perhaps even more than others, sense that something in their own mind is detached from them, something from which they feel estranged and to which, try as they may, they cannot find their way back. It is not to indulge in wholly remote speculation to think that the philosopher is disturbed by this state of affairs within himself. And it need not come as a complete surprise to discover that his ambivalence about the submerged part of his own mind, both his inner perception of its existence and his denial, finds expression in his philosophical work. In view of the fact that a sentence which expresses a philosophical theory does not, despite appearances, describe or assert the existence of a state of affairs, supposing that an unconscious thought finds expression in it helps us understand what holds the philosopher spellbound to his view and also keeps it at a comfortable distance from his curiosity.[26]

The philosophical view that a person's experiences are private to himself carries with it the idea of inaccessible mental contents, and this idea suggests a connection of the view with the inaccessible unconscious. The view itself says nothing about our pains and feelings, but appears to be a veiled way of expressing the perception of the existence of the unconscious and also of mitigating its disturbing quality. By his theory the philosopher reports his perception of the noumenon[27] within himself in an inverted way, in the form of a projection. He deflects the perception away from himself and onto others, who thus become externalized surrogates for his own unconscious. In the fantasy which accompanies this projection he also represents to himself his unconscious (his own mind in relation to other people), and in this way denies the fact that it is alien territory which lies beyond his reach. The words 'The experiences a person has are private to himself' gives rise to the false notion that they have descriptive content; and the spell they are able to cast can be explained only by supposing they *do* have unconscious content. One concealed thought which the words might very well express is that the unconscious is outside of us and also that the contents of the unconscious are really no different from the contents of the conscious part of the mind. The philosophical 'view' that experiences are private, or that no one's experiences are accessible to anyone else, seems to be a mask for stating that the unconscious exists but that its contents are conscious.[28] Several remarks Freud has made throw light on the way some people cope with the uneasy perception that part of their mind is a lost land that lies beyond the horizon of consciousness. They are especially revealing in the present connection. What he said needs to be quoted at some length and deserves to be read with care:

...the study of pathogenic repressions and of other phenomena which have still to be mentioned compelled psycho-analysis to take the concept of the 'unconscious' seriously. Psycho-analysis regarded everything mental as being in the first instance unconscious; the further quality of 'consciousness' might also be present, or again it might be absent. This of course provoked a denial from the philosophers, for whom 'conscious' and 'mental' were identical, and who protested that they could not conceive of such a monstrosity as the 'unconscious mental'. There was no help for it, however, and this idiosyncrasy of the philosophers could only be disregarded with a shrug. Experience (gained from patholo-gical material, of which the philosophers were ignorant) of the frequency and power of impulses of which one knew nothing directly and whose existence had to be inferred like some fact in the external world, left no alternative open. It could be pointed out, inci-dentally, that this was only treating one's own mental life as one had always treated other people's. One did not hesitate to ascribe mental processes to other people, although one had no immediate consciousness of them and could only infer them from their words and actions. But what held good for other people must be applicable to oneself. Anyone who tried to push the argument further and to conclude from it that one's own hidden processes belonged actually to a second *consciousness* would be faced with the concept of a consciousness of which one knew nothing, of an 'unconscious conscious-ness' – and this would scarcely be preferable to the assumption of an 'unconscious mental'.[29]

III

It is of special interest to apply the later Wittgenstein, or rather that part of Wittgenstein's later work which some philosophers find hard to fit into the continuity of his thought, to several philosophical statements in the *Trac-tatus*. Although it has become almost a commonplace, it is worth noting that our mind works at several different levels simultaneously and that what we are aware of at one level we can obliterate from our mind at another. John Wisdom has described the case of a person who under hypnosis saw a blank space wherever the definite article occurred on a page of print. The expla-nation of this curious state of affairs is that part of his mind blotted out what he saw with another part. This suppressing, or blotting out, mechanism seems

to be used by philosophers who read Wittgenstein's later writings with intellectual blindness to the revealing things he said about philosophy. It has to be
granted that the iconoclastic perceptions to which he gave expression do not
belong to the continuity of his *philosophical* thought, but rather, are breaks
in it. They are remarkable departures from conventional philosophy. This is,
perhaps, one reason why philosophers have been able to read Wittgenstein
with a Parmenidean eye that eliminates the unconventional things he said
about conventional philosophy.

As is known, Wittgenstein rejected the *Tractatus*, even though some of his
later thought is continuous with it; and this rejection can be best understood
if we look at it through the metaphilosophical spectacles he has given us.
Consider the following selection of statements.

What can be described can happen too, and what is excluded by the law of causality
cannot even be described. 6.362[30]

Belief in the causal nexus is a *superstition*. 5.136[31]

A necessity for one thing to happen because another has happened does not exist. There
is only *logical* necessity. 6.37[32]

Just as the only necessity that exists is *logical* necessity, so too the only impossibility
that exists is *logical* impossibility. 6.375.[33]

The impression these pronouncements make on us is that they advance
factual claims about what exists or does not exist and about the irrationality
of a common belief about how changes are brought about in things. Read in
conjunction with each other, the sentences, 'A necessity for one thing to
happen because another has happened does not exist' and 'Belief in the causal
nexus is a superstition', give rise to the idea that propositions like 'The light
must go out when the current is turned off' and 'A hummingbird cannot
carry off a hippopotamus' are all being declared false. They also suggest the
notion that a person who believes any of them to be true is holding a pre-
scientific belief, one which, like the belief that heavier bodies fall faster than
lighter ones, has been shown false by science. It is factual claims like these
that Wittgenstein seems to be giving expression to by his philosophical sentences. But if we pause to reflect on them and relate them to other of his
statements, we will realize that what he seems to be saying here is, to use a
favorite expression of F.H. Bradley, mere appearance.

If we dispel the mists generated by the empirical talk with which philosophical theories about causation are surrounded, we can see that they are not

empirical. A philosopher who declares that causation is a logically necessary connection between classes of occurrences is, obviously, not holding an empirical view about causation. Logical connections are not discovered by observing the behavior of things, in Wittgenstein's words in the *Notebooks* 1914–16, '. . . none of our experience is *a priori*'.[34] But neither does a philosopher give expression to an empirical claim who says that there is no causal nexus or says that causation is nothing more than constant conjunction. For he has ruled himself out from being able to say what it would be like for there to be a causal nexus, i.e., what it would be like for a change to take place in a thing by another thing acting on it. He has also ruled himself out from being able to say what else might supplement mere constant conjunction. Wittgenstein said in his *Notebooks* that whatever can be described at all could also be otherwise.[35] The implication of this is clear. The sentence 'a causal nexus (or productive causation) does not exist' expresses an empirical proposition only if the term 'causal nexus' has a use to describe something which, if it did exist, would make false what the sentence asserts. The sentence would express an empirical proposition only if matters could be other than it declares them to be. Since it does not describe what could be otherwise, it does not express a proposition about what is or is not the case.

We come to the same conclusion, if we bring in statement 6.362 above, 'What can be described can happen too, and what is excluded by the law of causality cannot be described.' The implication of these words is that the law of causality has no describable exception, which is to say that nothing can be described such that if it existed it would upset the law. It is clear that a philosopher who holds that the law of causality has no conceivable, or describable, exception implies that, as *he* construes the words which give expression to it, they do not state an empirical proposition. It may be useful to point out that to say that 'what is excluded by the law of causality cannot be described' is to imply that *nothing* is excluded by it; otherwise it would be possible to say what is excluded, i.e., it would be possible to say what exceptions to the law would be like. On this claim about the law of causality, taken at face value, it has the character of a tautology, which also excludes nothing. This in general is the character of an *a priori* true proposition: it excludes no describable state of affairs. Whether or not the words 'A necessity for one thing to happen because another has happened does not exist' are being used to express an *a priori* proposition, it is clear that they do not express one that is empirical, i.e., one that *excludes* a describable state of affairs. The sentence which is joined to these words makes this evident, namely, the sentence 'There is only *logical* necessity'.

A philosopher who states that 'the only necessity that exists is logical necessity', or that 'the only impossibility that exists is logical impossibility', has not arrived at his claim by an inductive procedure, or by anything comparable to an inductive procedure. The difference between his sentences and a sentence like 'The only horses that exist are wingless horses' stands out. One is empirical and excludes what *can* be described. The other is used by the philosopher in such a way as to preclude his describing possible exceptions. The term 'the only' does not function in his utterance in the way it functions in the nonphilosophical sentence. Its function in the philosophical sentence is more like the one it has in 'The only even prime number is two'.

Seeing this makes it natural to think that a philosopher who says, 'The only necessity that exists is logical necessity', has the idea that he is using terminology in the accepted way to express an *a priori* truth. It also makes it natural to think that he has the idea that the sentence, 'A necessity for one thing to happen because another has happened does not exist', expresses an *a priori* truth. Without again going into an explanation of the nature of logical necessity, it can be seen that if he had this idea he could be charged with being in error about actual usage, that is, with having the mistaken idea that the use of 'necessary' is no wider than the use of 'logically necessary'. Again without repeating reasons elaborated in similar connections elsewhere, the conclusion that he has this idea has to be rejected, and with it the notion that the philosopher labors under the idea that he is making *a priori* pronouncements.

The alternative conclusion, which invariably provokes emotional resistance but nevertheless has great explanatory power in its favor, is that instead of being stubbornly fixated to a mistaken idea about usage he is in some way retailoring usage, artificially contracting or even suppressing terminology. Instead of supposing him to think that the use of 'necessary' coincides with that of the term 'logically necessary' and that the use of 'impossible' coincides with that of 'logically impossible', this alternative requires us to suppose him to be contracting 'necessary' and 'impossible' into part only of their actual use, which he announces in the ontological mode of speech. Philosophers like to show contradictions or vicious infinite regresses in each other's views, and they do succeed sometimes in momentarily embarrassing each other, which is the sum total of what is achieved. In philosophy showing a contradiction of one sort or another in a view does not remove it from the collection of optional theories. The philosophical view that the only necessity which exists is logical necessity, or that only logical necessity is real necessity, construed as making a factual claim that equates the terms 'necessary' and 'logically necessary' is subject to the obvious objection that it implies an infinite regression, comparable to the regression G.E. Moore pointed out in the ethical view that an act's being right is identical with its being thought right.[36] The metaphilosophical[37] construction of the view, that represents it as belonging to the family of 'theories' of which conventionalism is a member, offers an explanation both of how an astute thinker can overlook an obvious difficulty, why he could be embarrassed by having it brought to his attention, and also why the difficulty does not remove the theory from the collection of optional philosophical theories.

Wittgenstein has said that a philosopher rejects a form of expression under the illusion that he is refuting a proposition about things. This remark applies to his philosophical theories about causation and necessity in the *Tractatus*. The artificial contraction of the use of 'necessity' and 'impossibility' and the suppression of causal verbs (there is no 'causal nexus') is conducted in the style of discourse which we use to talk about things and which we also use to express necessary propositions. Presented in this way the words create the false but vivid impression that indubitable theories about the existence or nature of phenomena are being stated. There can be no doubt that this illusion answers to an unrelinquished yearning in the depths of the minds of philosophers, who continue to think that they can obtain knowledge of things without taking the trouble to go to them. The recent words of an important metaphysical philosopher strengthen this idea: 'What philosophers have supposed they were doing was pursuing truth; they were thinking about the ultimate nature of things — more critically than the common man, more profoundly than the scientist, more disinterestedly and precisely than the theologian.'[38]

NOTES

* If one is afraid of the truth (as I am now), then one never apprehends the *whole* truth. Wittgenstein, *Notebooks 1914–1916*.

[1] See, for example, 6.53.

[2] *Philosophical Investigations*, p. 19.

[3] Ibid., p. 51.

[4] This phrasing of the idea follows the phrasing in The Yellow Book.

[5] Wittgenstein has brought out the same point with the help of his special use of the term 'grammar': 'The way you verify a proposition is part of its grammar. If I say all cardinal numbers have a certain property and all men in this room have hats, the grammar is seen to be different because the ways of verification are so different.' (Lecture notes, 1934–35, taken by Alice Ambrose). Wittgenstein undoubtedly uses here his word 'grammar' to refer to the difference in the use of the two statements, a difference in their 'logical grammar'.

[6] *An Analysis of Knowledge and Valuation*, p. 91.

[7] P. 53e.

[8] The phrase is taken from the *Tractatus Logico-Philosophicus*, 3.001.

[9] Wittgenstein's italics bring out the point.

[10] *Philosophical Papers*, p. 275.

[11] Ibid., p. 276.

[12] Ibid., p. 291.

[13] Ibid., p. 41.

[14] Notes of lectures, 1934–35, taken by Alice Ambrose. Compare the last sentence with the *Tractatus*, 6.126.

[15] *The Blue Book*, p. 57.

[16] For discussions of the view that propositions and the meanings of general words are abstract entities, see especially 'Understanding Philosophy' in my *Studies in Metaphilosophy*, and 'The Existence of Universals' in *The Structure of Metaphysics*.

[17] It should be noted that 'mean the same' does not apply to all sentences which translate into each other, for example, to equivalent nonsensical sentences in different languages. A person who insists that sentences which translate into each other *must* mean the same is stretching the expression 'sentences which mean the same' so as to give it the same range of application that 'sentences which translate into each other' has.

[18] *Remarks on the Foundations of Mathematics*, p. 136.

[19] *Philosophical Investigations*, p. 49.

[20] The Yellow Book.

[21] A.J. Ayer, *The Problem of Knowledge*, p. 202.

[22] G.H. von Wright, 'Deontic Logic and the Theory of Conditions', *Critica* **2**, p. 3.

[23] It is hardly necessary to call attention to the fact that the philosophical use of the word 'experience' does not correspond to its everyday use. We should not, for example, say of a person in pain that he is having an experience.

[24] *The Blue Book*, p. 46.

[25] *Philosophical Investigations*, p. 48. This is easily recognized as linguistic therapy for avoiding metaphysics.

[26] It also helps us understand why philosophers experience so much hostility to the notion of an unconscious part of the mind. The suspicion, which they may very well

harbour, that the sole ideational, non-verbal content of a philosophical theory is a cluster of unconscious thoughts would naturally provoke a strong reaction against the idea of an unconscious.

[27] Kant's notion of a noumenal mind behind the conscious self is easily recognized as referring to the unconscious.

[28] At a more superficial level the philosophical view would seem to express a self-revelation: a felt inability to empathize with others.

[29] *An Autobiographical Study*, pp. 55–6.

[30] My translation.

[31] D.F. Pears and B.F. McGuinness translation.

[32] C.K. Ogden translation.

[33] D.F. Pears and B.F. McGuinness translation.

[34] P. 80e.

[35] Ibid.

[36] *Ethics*, pp. 123–25.

[37] For an explanation of the term 'metaphilosophy' see my note in *Metaphilosophy* 1, p. 91.

[38] Brand Blanshard, review of *Philosophy and Illusion* in *Metaphilosophy* 1, p. 178.

CHAPTER IV

MYSTICAL AND LOGICAL CONTRADICTIONS

Wittgenstein has said that 'what we cannot speak about we must consign to silence'.[1] This utterance would seem to have the inviolate security of the tautology that what we cannot speak about we cannot speak about. Mystics, both western and eastern, however, are not deterred from violating, or trying to violate, it; they do not consign to silence what according to them is 'beyond expresssion'. They profess to experience something which is not only beyond the bounds of sense, but is also beyond the reach of language, while trying to express in both prose and poetry their special experience. One is inclined to say that mystics have undertaken a paradoxical task, one of trying to do what there is no trying to do. Just as there is no trying to open an open door, so there is no trying to communicate an experience which is *intrinsically* incommunicable.

One psychoanalyst has remarked[2] that to preserve the proper relationship between the master and the disciple, the initiate and the neophyte, the master must possess knowledge which is hidden from his disciple. In the case of the mystic it would seem that the distance and mystery that protect the master's knowledge is secure from penetration by others. The aura that surrounds the mystic encourages the idea that he has a privileged experience vouchsafed only to the elect; and we need not reject out of hand the idea that he does have an experience which does not occur in everyday awareness. But there is more to the mystic's talk than the need to hold his disciples and keep them at a respectful distance, as attention to the language he uses (his 'language game') shows. For the present we do not understand his apparent attempt to go against his own claim to having an incommunicable experience, his trying to disclose to others what he declares is not subject to disclosure. The thought which suggests itself is that the mystic's efforts are not in vain and that he does succeed in communicating something by means of the well-known self-contradictory language he uses, for example, 'the many which is neverthelesss one'. This means that his language is not to be understood in the normal way, i.e., as having self-contradictory meanings. Understood in the normal way it may truly be said to refer to the incommunicable, to that which is not susceptible of verbal description. It is possible, however, that it also has a superimposed use in which the self-contradictory forms of speech are made to refer to what is not beyond the reach of descriptive language. Undoubtedly the mind of the mystic, like that of anyone else, operates at more than one level, and he may be using language in more than one way and at more than one level of his mind. He may be using self-contradictory language in the normal way and creating mystification by the talk with which he surrounds it. And it is possible that at the same time he also is using it in a non-normal way, perhaps unconsciously, to convey certain ideas.

William James has described the achievement of the mystic as being 'the overcoming of all barriers between the individual and the Absolute':[3] 'In mystic states we become one with the Absolute. This is the everlasting and triumphant mystic tradition, hardly altered by differences of climate, culture, or creed. In Hinduism, in Neo-Platonism, in Sufism, in Christian mysticism we find the same recurring note, so that there is about mystic utterances an eternal unanimity which ought to make the critic stop and think.' The following are some of the characteristics attributed to the mystical experience. It entails, for one thing, a state of mind described by the term 'unitary consciousness', a form of awareness which is said to transcend all multiplicity and difference. Timon, in a satiric poem, makes Xenophanes say, 'Wherever I turn my mind, everything resolves itself into a single Unity'.[4] Regardless of its satiric intent, Timon's description of the way things presented themselves to Xenophanes is echoed in the writings of mystics and some of the great metaphysicians. For another thing, the mystical experience is of something that can neither be perceived nor conceived. For a third thing, it is of something that is neither in space nor in time. And for still another thing, the object of a mystical experience can be characterized only in self-contradictory terms. Its nature can be conveyed only in paradoxes: it is that which exists while not existing, an emptiness which is a plenum, an immobility which moves, a nothingness which is yet something.

It is important in connection with such talk to call attention to what Freud has said about the id: 'The logical laws of thought do not apply to the id, and this is true above all of the law of contradiction. Contrary impulses exist side by side without cancelling each other out or diminishing each other; at the most they converge to form a compromise.'[5] These words appear to attribute to the id what mystics attribute to the object of their experience: a self-contradictory nature, Freud appears to imply, by his words, that psychological phenomena which violate the 'law of contradiction', i.e., self-contradictory phenomena, exist in the unconscious. But Freud was not a mystic with respect to primary process psychology, although his language at times suggests this. A sober reading of his words, which requires bridling a tendency to be a semantic Don Quixote, will show that he is not using the term 'law of contradiction', whether by design or by mistake, in the way it is used in formal logic and mathematics. In his manner of speaking, a violation of the law of contradiction occurs when contrary impulses exist side by side; and, plainly, to say that contrary, or opposing, impulses exist side by side is neither to say nor to imply that one and the same impulse has contradictory properties. It is plain, also, that in stating that contrary impulses may converge to form compromises, it is no part of his intention to imply that two conflicting impulses may converge to become a self-contradictory impulse, an impulse which is logically impossible. To make a general observation, it seems that there is a tendency in those who are outsiders to formal logic and mathematics to equate 'conflicting' with 'contradictory'. The result is confusion and misunderstanding.

Many people feel baffled by the mystic's account of his experience, and one reaction is simply to dismiss it as aberrated talk. It is quite easy to single out a number of attitudes towards his account. Some people have the idea that the mystic does in fact have an experience of radiant illumination but which they think is subjective, that is, not an awareness of a reality. Others, like W.T. Stace, believe that the mystic is in touch with some sort of being. Thus Stace has asked whether the mystic's words refer to 'a beautiful dream',[6] and he produces what he takes to be evidence for thinking it is not just a dream and that it does bring him into contact with an actual being. Still others, like A.J. Ayer, think that the talk of the mystic is literal nonsense, that when he speaks of a unitary consciousness which transcends all multiplicity and difference, and which is characterizable only in self-contradictory terms, he in indulging in gibberish. According to this view-point the mystic's talk is linguistically bogus and has no use to refer to or to describe an experience, whether subjective or objective.

We are familiar with disagreements over whether certain claimed perceptions, individual or group, are delusive, on a footing with dreams and hallucinations, or whether they are of actual objects, as in my present perception of the sheet of paper on which I am writing. Such disputes are empirical, subject to a variety of tests; and their investigation does not fall within the field of a philosopher's special competence. Nor is it of more interest or concern to him than it is to nonphilosophers. But the mystic's claim that the content of his experience, regardless of whether it is objective or subjective, is characterized by contradictory predicates, one may naturally think, is a proper subject of investigation for him. Philosophers, as the literature would seem abundantly to show, are especially skilled at discovering and producing contradictions in unsuspected places, although it has to be said immediately that these contradictions themselves become subject to dispute. Thus, hardly anyone except a philosopher thinks that a contradiction lies buried in the idea of a thing being in motion or in the idea of there being a multiplicity of things. As is well known, philosophers like Zeno and F.H. Bradley have given reasons which they assert show contradictions in these and many other concepts; and there would seem to be justification for the idea that they are specially suited to evaluate cogently the mystic's nonempirical claim that the reality he is in touch with has a self-contradictory nature. In the present chapter I shall limit myself to an investigation of this claim.

It is important to point out a number of differences between the claims of philosophers like Zeno and Bradley and those made by mystics. These philosophers profess to demonstrate contradictions in concepts which are commonly taken to be exemplified in the real world. And they have also attempted to prove that reality is a pure unity which transcends variety and multiplicity. Mystics, on the other hand, claim to experience objects which are exemplifications of self-contradictory concepts, and they also report experiencing the transcendent unity, or the undivided One. Parenthetically, it would seem to add a paradox to the collection of the mystic's paradoxes to remark that his reported experience is of an undifferentiated unity which is yet differentiated by predicates, albeit incompatible ones. A curious difference now emerges. Whereas the mystic reports experiencing self-contradictory realities, philosophers like Zeno and Bradley maintain that establishing a contradiction in a concept shows that no reality can answer to it. And the immediate consequence with regard to the mystic's experience is that it is subjective. His experience does not bring him into touch with a reality. Its content, to use Bradley's expression, is mere appearance. Bradley's answer to Stace's question whether the mystical experience, in which the self-contradictory reveals itself, is no more than a beautiful dream, would have to be that it is. Another consequence is that an experience which is confined to the privileged few is thrown open to the general public. Everyone perceives the self-contradictory. This can easily be seen from the following consideration.

An appearance, ϕ, of a self-contradictory state of affairs E (which because if is self-contradictory cannot exist, or be real) will itself have to be self-contradictory. If ϕ were internally consistent it would be an appearance or representation of a state of affairs which could exist. It would be a picture of a possible reality rather than of a self-contradictory one. Hence if motion were self-contradictory, the appearance of something being in motion must also be self-contradictory. And if time were self-contradictory, appearances of succession and change would be self-contradictory. Now, the appearances exist and are perceived by everyone. Consequently everyone, and not only mystics, have experiences whose contents are self-contradictory. It should be pointed out that Blanshard has allowed that if being self-contradictory prevents a reality from existing it also prevents the appearance of that reality from existing, and that he would take his stand with reason and deny the existence of the appearances.[7] Without going into the matter, this would seem to be a way of arriving at Gorgias' proposition that nothing exists. In philosophy, unlike any other subject which professes to investigate the world, there is no feeling of absurdity about the picture of a philosopher saying to his class, 'Gentlemen, this morning I shall try to demonstrate to you that nothing exists'.

The mystic reports having an experience of a reality which moves while not being in motion, and exists while not existing. This is enough to dumbfound the layman. But when he consults a philosopher like Bradley, Zeno, Parmendies, and others, who are knowledgeable in the ways of contradiction, he is told that no reality can be self-contradictory but that sensible appearances commonly taken to be real are self-contradictory. He is informed, to his astonishment, that like the mystic he perceives contradictions, that, for example, when something looks to him to be in motion, he does actually perceive the appearance of an unending process which yet comes to an end. From another highly regarded philosopher he learns that what seems to him to be happening cannot seem to be happening because the appearance of a thing being in motion does not exist. If he goes on to still another philosopher who has made a special study of contradictions, he is introduced to still another baffling opinion. Thus, Kant has attempted to show that certain concepts involve special contradictions called antinomies, which resemble what in formal logic are called logical paradoxes, for example, the paradox implied by the idea that there is a class of all classes. Kant professed to show that a composite substance is made up of simple parts and that it cannot be, that the world had a beginning in time and could not have, that it comes to an end in space and that it is an infinite magnitude, etc.

As an aside, it is of interest to note that historically these theses and antitheses were regarded as rival positions between which a decision could in time be reached. Kant's claim is that a decision in favour of either one is impossible. His important idea with regard to what the presence of an antinomy shows is that, in a metaphor, it is a closed door to the world of noumena, the reality which is not in space and time. His well-known view is that noumenal reality is unknowable, not that it is unknowable by ordinary mortals and knowable by mystics, but that it is unknowable by anyone: his idea was that the existence of antinomies implies that whenever we try to think of noumenal reality we become enmeshed in contradictions of a sort which *prevent* us from experiencing it. They are not a medium through which it reveals itself.

To someone who is spectator to this conflict of opinions the situation is bewildering, to say the least. The mystic, with unquestionable sincerity, states that he is conscious of a self-contradictory reality that lies beyond space and time. Some philosophers are led by chains of reasoning to the view that whatever is characterized by contradictions is unreal, which would seem to imply that the mystical experience is not essentially different from that of the common run of people who perceive the appearances. And Kantian philosophers embrace the view that our minds are barred from the realm beyond space and time, i.e., that which transcends multiplicity and difference, by the existence of certain contradictions. To this array of views a still further view needs to be added. This is that self-contradictory expressions have no use in the language to describe or in any way refer to anything, either realities, or appearances, or barriers. The consequence of this last position is that anyone who asserts that he experiences a reality that is characterized by self-contradictory attributes, or asserts that everyone in his everyday experience perceives self-contradictory phenomena, or maintains that antinomies are insurmountable obstacles to our having a higher-order experience, is not using language to tell us anything about experience. All that he succeeds in doing is to misstate, or misdescribe, the functioning of language. A person who is audience to this 'prodigious diversity'[8] of apparently incompatible claims may be forgiven if he is perplexed and would like to rise above the multiplicity and reach some sort of understanding of the situation. The notion of self-contradiction is central to the various claims, both mystical and philosphical; and one thing it is important to get clear about is the way self-contradiction works, if we are to reach an insight into the situation. To imitate one of the last things Wittgenstein says in the *Tractatus*, one must transcend these claims in order to see them aright. And this can only be done by improving our understanding about how *logical* self-contradiction works and whether the contradictions of the philosopher and the mystic are, as they do appear to be, logical contradictions.

The idea that self-contradictory entities exist, regardless of whether they are objective or subjective, a beautiful reality or a beautiful dream, the idea, in other words, that there is or could be something answering to a self-contradictory concept, implies that a logical impossibility is really a factual impossibility, whether physical or psychological. A proposition which declares an occurrence or a state of affairs to be factually impossible is empirical and is open in principle, if not in fact, to falsification. A physical impossibility, e.g., the impossibility of water boiling at a temperature of 70° below zero, is linked with the conceivability of the opposite. What is physically imposssible could be conceived of as taking place. We might call an occurrence which goes against a physical immutability a miracle, but there is no question about the conceivability of its occurring. There is an aura of the miraculous and the occult about the claimed experience of the mystic, but the occurrence of anything answering to a self-contradictory concept will have to be a *logical* miracle, one which there is no conceiving, by either mystic or philosopher. For a logical miracle would entail the instantiation of a logical impossibility, which differs from a factual impossibility in not having a conceivable opposite. To put forward a claim which implies that a logical impossibility is a factual imposssibility and, thus, that a proposition which states a logical imposssibility is empirical, is to make a claim that implies that a logical impossibility is not a logical impossibility. The only conclusion would seem to be that the claimed experience is itself logically impossible and the claim not genuine. The mystic and the philosopher cannot be charged with charlatanism or insincerity. We may arrive at some understanding of what has happened by considering a consequence of the proposition that there are self-contradictory entities.

Whatever is open to being conceived, even though a supermind would be required to conceive it, *could be*; and if a logical impossibility were conceivable, it could exist. What prevents a logically impossible, and therefore self-contradictory, state of affairs from existing also prevents it from being conceived. The mystic, of course, maintains that the object of his awareness cannot be conceived. The implication of this is something the mystic would not be willing to accept, namely, that he attaches no meaning whatever to the words he uses to refer to the object of his claimed experience, and thus that he is talking literal nonsense. Words ostensibly used to refer to what is not conceived are words to which no literal meaning is attached. Spinoza said that not even God can bring it about that from a cause no effect shall follow, and he might have added that not even God can conceive of a cause which has no effect. Anything that an omnipotent Being can conceive he can bring about. But there is no conceiving or imagining something causing a change while having no effects, because there is no conceiving a logically impossible state of affairs — not by the mystic nor by God. Wittgenstein asserted that 'if a god creates a world in which certain propositions are true, he creates thereby also a world in which all propositions consequent on them are true. And similarly he could not create a world in which the proposition 'p' is true without creating all its objects'.[9] The reason why he could not is that 'we could not *say* what an 'illogical' world would look like'.[10] Not being able to *say* what an illogical world would look like is not due to a shortcoming in our vocabulary, which could be made good. It is due to what might be called a built-in fact of language, the fact that an expression which stands for a logically impossible concept has no use to describe or refer to an object, an event, an appearance, or a barrier to an unexplored realm.

Consider the sentence, 'There is no greatest prime number'. The proposition it expresses, unlike the proposition expressed by the sentence 'There is no gas which cools when compressed', is *a priori* true and has no theoretical falsification: there is no hypothetical number which could make it false.

Anyone who stated that he sometimes conceived the greatest prime number would be dismissed by a mathematician as talking literal nonsense. The proposition is demonstrated by showing that its denial, namely, that there is a greatest prime number, implies a contradiction. And what this in turn shows is that the phrase 'greatest prime number' in the sentence 'There is no greatest prime number' has a self-contradictory meaning and has no use in the language to refer to a number. In general, having a self-contradictory meaning, which implies standing for a logically impossible concept, prevents an expression from having a characterizing use. It should be pointed out immediately that the statement that every expression which has a self-contradictory meaning lacks a referential use is not an inductive generalization based on a canvas of self-contradictory meanings. There are not two sorts of expressions which have self-contradictory meanings, those which lack use and those which *might*, nevertheless, have a use in the language, such that which sort a given expression is requires additional investigation. If an expression does not stand for a logically impossible concept, it does not have a self-contradictory meaning; and hence, by *modus tollens*, having a self-contradictory meaning *entails* lacking a use to refer to or characterize something. To be sure, an expression may have the form of a self-contradiction, without having a self-contradictory meaning. If, for example, someone says 'Yes and No' in answer to the question, 'Is it raining?', or in answer to the question, 'Is so-and-so bald?', he is conveying factual information about an in-between condition of the weather or so-and-so's head. Similarly, a noncontradictory meaning is conveyed by the grammatically contradictory sentence, 'For three thousand years time stood still here'.[11]

It is instructive at this point to compare the interpretation Kant placed on his antinomies and the interpretation Bertrand Russell placed on his class of classes paradox. Kant professed to have demonstrated that the thesis, e.g., *the world is a finite magnitude*, is bound up in a vicious circle with its antithesis: it is implied by its own denial and it implies its own denial. This showed, he thought, that our experience was confined to the sensible manifold and that our minds could not rise above it and explore the realm of being which transcends space and time. Russell showed that the notion of a class of all classes also involves an antinomy, namely, the class of all classes which are not elements of themselves is a class which both must be and cannot be an element of itself. The interpretation which he placed on the class of classes paradox is that it shows an expression to be 'meaningless'.[12] It shows, according to him, that the term 'class of all classes' does not have a use to designate a class, and not that it designates a class which is revealed only in special moments of illumination. If Kant has indeed demonstrated a vicious circle contradiction in the proposition that the world is a finite magnitude, then what he has shown is that any expression standing for the concept of *the world*, or of *the world as a whole*, is devoid of referential sense: formulated in the English language, it shows[13] that the phrase 'the world as a whole' has a self-contradictory meaning, such that in that meaning it has no use to designate an object, any more than 'class of all classes' has a use to designate a class or 'greatest prime number' has a use to designate a number.

The same remarks apply to the philosophical positions of Zeno and Bradley. If they *have* demonstrated by their lines of reasoning that the concepts of change, multiplicity, motion, qualitative diversity, etc. imply contradictions, then the terms whose meaning these concepts are have no use to describe or characterize or in any way to allude to a state of affairs. Unlike the sentence, 'There are no winged zebras', and like the sentence, 'There are no wimbling toves', the sentence, 'There are no moving trains', makes no declaration about what does not exist. Furthermore, if the expression 'moving train' has a self-contradictory meaning, then so does the expression 'sensible appearance of a train being in motion'. It has no more use to refer to a sensible appearance than does the expression 'sensible appearance of a wimbling tove'. If the demonstrations of their putative claims are correct, then Zeno and Bradley have shown nothing about the nature or existence of things or appearances. Their views are neither about what there is nor about what there appears to be. They are instead about the use of classes of expressions and are to the effect that they have no correct use in everyday speech. The notion that this is what the views come to, although not expressed in the form of speech in which words are mentioned, has led Norman Malcolm to maintain that all such views can 'be seen to be false *in advance* of an examination of the arguments adduced in support of them'.[14] According to him, a philosopher is made blind by persuasive reasoning to what is right before his eyes, namely, obvious facts of correct speech. Regardless of whether one finds persuasive this account of philosophical claims that the meanings of certain expressions in ordinary language embody contradictions, and thus are logically impossible concepts, the secure point is that an expression whose meaning implies a contradiction or a logical impossibility has no use in that meaning to designate an obstacle, a reality, or an appearance. Taken at face value, the mystic's statement does not report the content of an experience, any more than do the philosophers' claims to have brought to light a barrier or the unreality of things.

There is no reason to doubt that the mystic actually has a profoundly moving experience which he tries to communicate, in part, by using baffling, paradoxical language, language which makes some people shake their heads incredulously and makes others lose interest completely. If we give up the idea that he is trying to describe to us a logically impossible experience, one which is beyond all expression, the only correct inference to draw is that his use of self-contradictory expressions does not correspond to the function they normally have. His use of self-contradictory terminology is a departure from its use in ordinary language, as well as in mathematics and logic: he turns into designations expressions which have no use to refer to anything. The problem then becomes one of determining why he chooses self-contradictory expressions for this purpose, what motive using them serves. One possible motive lies on the surface: this is to create a feeling of mystery and awe in others, comparable, we may think, to that experienced by participants in ancient mystery rites.

In the English language we sometimes encounter a word which has two opposing senses: the word 'cleave' has the double meaning of split asunder and adhere to. And it is of special interest in the present connection to learn that in many of the oldest known languages the same word or symbol may have antithetical meanings, requiring special devices to indicate which meaning is intended. Karl Abel, who investigated the occurrence of primal words which simultaneously denoted a thing and also the opposite of that thing, observed that 'Man has not been able to acquire even his oldest and simplest conceptions otherwise than in contrast with their opposite; he only gradually learnt to separate the two sides of the antithesis and think of the one without conscious comparison with the other.'[15] In other words, a contrast which was originally imbedded in the meaning of one word became later a contrast between the meanings of two words which could be used separately and needed no indicating devices. So to speak, a verbal hermaphrodite separates into a pair of Siamese twins, one of which is male and the other female: they are inseparable but individual.

In dreams, which use archaic modes of expression, there sometimes occurs representation of a thing by its opposite, so that within the context of the dream itself it may be impossible to determine whether an element which is capable of an opposite is to be taken as itself or as its opposite. Freud was led to make the observation that, 'Dreams show a special tendency to reduce opposites to a unity or to represent them as one thing'.[16] The phrase 'reducing opposites to a unity' has a mystical ring to it, but it refers to nothing more mystical than the use of a symbol to represent both of a pair of opposites. And the phrase 'representing opposites [e.g., being dressed and being naked] as one thing' refers to nothing more mystical than the use of opposites to stand for each other. Abel said that the ancient Egyptians '. . .must at least have had sufficient sense not to regard a thing as at one and the same time itself and its opposite',[17] and all he could have meant by these cryptic words was that they did not think that self-contradictory expressions had a use in their language to refer to or describe states of affairs. Nor did Freud wish to imply that there are self-contradictory representations in dreams. When a witty philosopher argues that we are all really naked, because under our clothes we do not wear clothes, he does not imply that a person who is dressed is yet undressed. Instead he is in a concealed way telling us that nakedness can be represented by its opposite, being clothed. We now are in a better position to understand a mystic who says, 'Black does not cease to be black nor white white. But black is white and white is black. The opposites coincide without ceasing to be what they are in themselves'.[18] He is unconsciously harking back to archaic usage which assigns antithetical senses to the same word. To paraphrase Freud, mystics show a special tendency to reduce opposites to a unity, which is to say that they have a tendency, or wish, to assign antithetical senses to the same word. This may be so because they are more under the domination of unconscious processes in their minds than are ordinary people and that their waking life is more permeated by dream life than is ours.

It is of interest to investigate another possible determinant of the mystic's choice of contradictory forms of speech. In one of his discussions of the mystical paradoxes Stace uses language which suggests, very likely without conscious intention, that they give expression to ambivalence about the object with which the mystic believes himself to be in contact. He wrote, 'Their general character is this: that whatever is affirmed of God [or Nirvana, the One, etc.] must be, at the same time and in the same breath, categorically denied. Whatever is said of the Divine Being, the opposite, the contradictory, must also be said.'[19] Self-contradictory formulations, which are composed of conjunctions of symbols with symbols which are or imply their contradictories are well suited to serve for the expression of ambivalence. What one side of the formulation affirms, the other side negates. This makes it psychologically natural to use the two sides to express or to indicate conflicting feelings, such feelings as love and hate, which when directed to the same thing create ambivalence, a tension of opposites, to put it in Heraclitean language. It is natural, for example, to equate movement with life and immobility with death. This makes it possible for the paradox of the One which is stationary while overtaking 'those who run' to give hidden expression to a condition of inner ambivalence, a conflict between a death-wish and the counter-wish which are projected onto the same object. Again, Nothing may, amongst other things, be identified with the negligible and the unimportant, and Being with the exalted and the important. For someone who unconsciously makes these associations, the paradox of that which has being while yet it is nothing is a possible way of giving expression to opposing tendencies towards the same object, one tendency to deprecate and despise and the other to elevate and exalt. One way of dealing with what might be called primary ambivalence is to split the image into two images, one of which is loved, the other hated: shining Baldur and malevolent Loki, the radiant Ormazd and the evil, dark Akriman.

Eckhart's statement, 'Black is white and white black', despite the fact that it has no use in actual language to characterize the colors black and white, may become an unconscious way of referring to an inner state of affairs, a state in which, so to speak, black and white are in conflict with each other. To black it is natural to associate malevolence and evil, against which we may, without feelings of guilt, turn our hatred; and to white it is natural to associate purity and goodness, to which unmixed love can be directed. In Eckhart's case it would seem that the splitting mechanism was not employed, either that, or else that the divided images at times coalesced for him, with the result that the object of his mystical experience was a demon God, adored and desired and hated and feared simultaneously.

ADDENDUM
A MAXIM OF MODAL LOGIC

In connection with what has been said here about the difference between logical and factual impossibility, it needs to be pointed out that some philosophical logicians lay down a maxim which might be used in defense of the mystic's claim to apprehend a logically impossible object or occurrence. The maxim, found in systems of modal logic, is that a logically impossible proposition, one which is necessarily false, is also false as a matter of fact:

$$(\alpha) \qquad \sim \Diamond p \rightarrow \sim p.$$

It might be argued that if the factual falsity of p is validly inferrable from its logical impossibility, so also is the factual impossibility ($\sim \Diamond p \rightarrow p$ is factually impossible). Letting $\blacklozenge p = p$ is factually possible, we have:

$$\text{If} \sim \Diamond p \rightarrow \sim p, \text{then} \sim \Diamond p \rightarrow \sim \blacklozenge p.$$

Thus, a modal logician who held (α) would be entitled to hold

$$(\beta) \qquad \sim \Diamond p \rightarrow \sim \blacklozenge p.$$

And with (β) he would seem to be in a position to protect the mystic's claim. For that which is factually impossible, such as cooling a gas by compressing it, is conceivable, and could be made to occur by Divine intervention. Hence, by (β), what is logically impossible is conceivable, although perhaps only the mystic and God are capable of conceiving it. In telling us what we cannot comprehend, because it is characterized by contradictory predicates, the mystic comprehends what eludes lesser minds, and, for all we know, is aware of a self-contradictory reality.

It is not easy to think of logicians as being allies of mystics, and it may be that they are so in linguistic appearance only. Hume stated that from what is necessary nothing contingent follows, and this is a secure thesis which no logician rejects. A proposition which has its truth-value by logical necessity, as against a proposition which has its truth-value contingently, is such that its actual truth-value is identical with all of its possible truth-values. This is because it has only one possible truth-value. A proposition which has its truth-value contingently is capable of either of two possible truth-values, so that its actual truth-value cannot coincide with all of its possible truth-values. These comments are, of course, merely elaborations of the ideas of necessary and contingent propositions, and the point of making them is to put Hume's thesis in an improved light. His thesis comes to saying that the actual truth-value of a proposition which has two theoretically possible truth-values is not entailed by and cannot be calculated from a proposition which has only one possible truth-value.

It follows directly that a true proposition which declares something to be factually impossible, and thus has two possible truth-values, is not entailed by nor can it be calculated from a logically impossible proposition, which has only one possible truth-value. It thus follows that from the fact that a proposition, p, is logically impossible it cannot be validly inferred that p is physically impossible. The same line of reasoning holds good for the statement that the matter of fact falsity of a proposition is entailed by its logical impossibility, where being false as a matter of fact, as against being false by logical necessity, implies having more than one possible truth-value. All this lies plain on the surface, and there is only one explanation of what a logician who asserts (α) is doing, if we wish to avoid charging him with a gross blunder to which he remains attached. The explanation is that he is using the sentence, in whatever language, in which he formulates (α), not to give expression to a putative entailment between necessary and contingent propositions, but to give oblique expression to a linguistic fact. This is the fact that in the language of mathematics and logic expressions to the effect that so-and-so is necessarily true or that so-and-so is impossible are used interchangeably with expressions that so-and-so is true or that so-and-so is false. Unlike its use in ordinary language, the use of the factual idiom is not different from the use of the modal idiom, in certain respects. Thus, Bertrand Russell's rejection of C.I. Lewis' notion of strict implication, which involves the modal notion of necessity, in favor of so-called material implication, p ⊃ q, which is a relation expressed by 'either p is false or q is true', is to be construed as discounting a certain form of words, the modal idiom, as not needed in mathematics. Russell's own words make this quite plain. He has written: 'I maintain that, whether or not there be such a relation as Lewis speaks of, it is in any case one that mathematics does not need, and therefore one that, on general grounds of economy, ought not to be admitted into our apparatus of fundamental notions.'[20] Russell refers to notions rather than to forms of language, but it is more natural for philosophers to speak of ideas and notions than of words. If we keep in mind Wittgenstein's observation about 'the confusion that pervades all philosophy', which consists of thinking a philosophical problem to be about 'a fact of the world instead of a matter of expression',[21] what comes through in the Russell passage is that mathematics has no need for modal terminology *in addition to* the factual language it employs. This is so because *in mathematics* the two forms of speech mean the same. The economy which Russell advocates is an economy of notation, not an economy of notions. And a philosophical logician who adopts the maxim that a logically impossible proposition is false as a matter of fact does nothing more remarkable than resist Russell's notational economy and come out in favor of retaining both forms of speech in logic and mathematics. Only in appearance can his maxim be put into the service of mysticism.

The mystical view that contradictory predicates characterize a special Being (and also the view that sensible appearances are self-contradictory) may be described as a structure consisting of three interacting parts, two of which are hidden from awareness. Without going into detail, it is not difficult to see that such an expression as 'the emptiness which is everywhere full' presents a grammatical innovation, namely, the artificial classification of self-contradictory predicates with predicates that have a characterizing use, i.e., with predicates denoting attributes. This classification is, to use Wittgenstein's word, idle in the language: the predicates are not actually given a use to refer to features of things. Without being aware of what he is doing, the mystic turns a self-contradictory grammatical predicate into a holiday semantic predicate, whose function, at one level of the mind, is to produce a striking illusion and, at another level of the mind, to express thoughts which are detached from consciousness.

NOTES

[1] *Tractatus Logico-Philosophicus*, 7. D.F. Pears and B.F. McGuinness translation.
[2] Dr Charles Brenner, in a conversation.
[3] See his chapter on mysticism in *The Varieties of Religious Experience*.
[4] Quoted in Theodor Gomperz' *The Greek Thinkers*, Vol. I, p. 159.
[5] *New Introductory Lectures on Psycho-analysis*, newly translated and edited by James Strachey, pp. 73–4.
[6] 'Mysticism and Human Reason', Reicker Memorial Lecture, no. 1, *University of Arizona Bulletin Series*, p. 11.

[7] See 'In Defense of Metaphysics', in W.E. Kennick and Morris Lazerowitz (eds.), *Metaphysics. Readings and Reappraisals*, p. 348.

[8] Hume's phrase.

[9] *Tractatus Logico-Philosophicus*, 5.123. Ogden translation.

[10] Ibid., 3.031. Pears and McGuinness translation.

[11] Gleaned from Paul Herrmann's *Conquest by Man.*

[12] *Principia Mathematica*, second edition, p. 37.

[13] This way of saying what the consideration shows is a short-cut which rests on a number of partial explanations I have given elsewhere. For an important part of the complete explanation the reader should return to Chapter III, 'Necessity and Language', Section I.

[14] *Knowledge and Certainty*, p. 181.

[15] Quoted from Karl Abel by S. Freud in 'The Antithetical Sense of Primal Words', *Collected Works*, Vol. IV, p. 187.

[16] S. Freud, op. cit., p. 184.

[17] Karl Abel, op. cit., p. 186.

[18] Quoted by W.T. Stace in *Mysticism and Philosophy*, p. 65.

[19] Op. cit., p. 12.

[20] *Introduction to Mathematical Philosophy*, p. 154.

[21] The Yellow Book, quoted by Alice Ambrose in 'Philosophy, Language, and Illusion', *Psychoanalysis and Philosophy* (eds. Charles Hanly and Morris Lazerowitz). See pp. 22, 25.

CHAPTER V

MOORE'S ONTOLOGICAL PROGRAM

In his lectures on *Some Main Problems of Philosophy*, which Moore gave fifteen years before his 'Defence of Common Sense', he defines the task of philosophy in the following passage:

...it seems to me that the most important and interesting thing which philosophers have tried to do is no less than this; namely: To give a general description of the *whole* of the Universe, mentioning all the most important kinds of things which we *know* to be in it, considering how far it is likely that there are in it important kinds of things which we do not absolutely *know* to be in it, and also considering the most important ways in which these various kinds of things are related to one another. I will call all this, for short, 'Giving a general description of the *whole* Universe', and hence will say that the first and most important problem of philosophy is: To give a general description of the *whole* Universe. Many philosophers (though by no means all) have, I think, certainly tried to give such a description: and the very different descriptions which different philosophers have given are, I think, among the most important differences between them. And the problem is, it seems to me, plainly one which is peculiar to philosophy. There is no other science which tries to say: Such and such kinds of things are the *only* kinds of things that there are in the Universe, or which we know to be in it.[1]

It is to be noticed that Moore conceives philosophy to be one of the sciences. Many, and perhaps all, philosophers have this idea. Quine, for instance, states that philosophy is 'continuous with science',[2] and also that it is 'an aspect of science';[2] and he has the idea that 'what has drawn most theoretical physicists to physics is a philosophical quest for the inner nature of reality'.[2] The implication of Quine's words is that philosophy gives one what ordinary physical science cannot give, that it is a science which goes deeper than physics. Freud somewhere says that what we get from science we cannot get elsewhere, and Quine implies that a physicist can get from what we might call 'philosophical physics' what he cannot get from physics itself — information about the inner nature of things. This calls to mind Moore's remarks about different philosophical views as to what it is for a hen to lay an egg. According to some philosophers the analysis of the concept *material thing* shows an egg (and also a hen) to be a colony of monads. According to other philosophers analysis shows it to be a cluster of sense contents, etc. The important thing to notice about these philosophical theories, which are the results of concept-analysis, is that they lie beyond the bounds of possible investigation by physical science, that *in principle* they are not subject to investigation by observation or by experiment. Nevertheless, they do seem to be propositions about the inner constitution of things, about what things really are. This makes it natural to think that philosophy is continuous with science, and that the technique of analysis enables us to probe deeper into things than does either experimentation or observation, however refined the scientific aids to our senses may be. One of Wittgenstein's well-known remarks is that a philosophical sentence is like an engine idling, which is to say that it has no use to convey information. By contrast, Moore and Quine seem to think that the use of philosophical sentences is to convey deep information about things, as well as about other matters.

Moore's words, in which he defines the task of philosophy, describe what appears to be an ontological program, a program to determine in a general way what the contents of the cosmos are. Many of these are given by what have been called truisms of common sense. To cite two of these truisms: There exists at present a living human body, which is my body; there have also been many other things, having shape and size in three dimensions, from which it has been at various distances.[3] These are truths which obviously are not articles of common knowledge: not everyone knew that Moore existed. However, they imply, or appear to imply, basic commonsense truths which are known to everyone: space exists; material things exist. We may guess that Moore considers these two propositions truisms of common sense, not only because they follow from his initial particular truths but also because the man in the street, if asked whether space and things exist, would unhesitatingly say 'Certainly'. As an aside, we can imagine Wittgenstein adding that a nonphilosopher would do so with a bewildered look on his face. What I have in an early paper called 'Moore's Paradox' is pertinent at this point. 'It is a strange thing,' Moore observed, 'that philosophers should have been able to hold sincerely, as part of their philosophical creed, propositions inconsistent with what they themselves *knew* to be true.'[4] The plain implication of his words is that philosophers sincerely hold views they know to be false. If this is the case, it certainly is strange and remarkable, something which must bemuse anyone who pauses to reflect on it.

Philosophers whose views collide with common sense can hardly be expected to assent to Moore's charge. And to be sure, some philosophers have declared that Moore's paradox begs the question, that it assumes what is in dispute. If confronted with the paradox, a metaphysical philosopher like F.H. Bradley would no doubt say that far from knowing his views to be false, he knows them to be true and the so-called truisms of common sense to be mere commonsense superstitions. It will be remembered that Wittgenstein called the commonsense belief in causation a superstition (*Tractatus*, 5.136). The question arises as to how we are to decide between the claims of philosophical common sense and the reasoned views of metaphysical philosophers.

It can be seen immediately that one thing which cannot enter into our adjudicating between the rival positions is the testimony of the senses. For it goes without saying that all parties to the dispute see the same, hear the same, etc.; there is no question of testing anyone's eyes or hearing. There is no actual question that a philosopher who maintains that material things do not exist sees what a philosopher sees who maintains that they do exist. When Diogenes, in refutation of Zeno's proposition that motion does not exist, walked before the assembled scholars, there was no important difference in what was seen by those present. This means that sight could play no role in the settling of the dispute. Parmenides, it will be remembered, urged us not to heed the blind eye and to disregard the testimony it presents. Moore, on the other hand, might be construed as taking the counter position and urging us to heed our eyes and take into account their testimony. The curious thing about these two rival positions is that the adequacy of no one's sight is brought into question. No one would think of calling in an oculist. In urging the use of one's eyes a philosopher does not imply that a Parmenidean's vision is faulty and that he fails to see what the philosopher sees. The strangeness of urging the use of one's eyes on a person who we know sees what we see hardly needs to be elaborated on. We might restate the rival positions in this way. One declares that the testimony of the senses should be disregarded, the other recommends that we take it into account in our investigation of the world. What cannot fail to become plain to anyone who frees himself from embroilment in the debate is that the testimony of the senses cannot be used to resolve the disagreement. A philosophical debate in which the use of the senses is in question cannot be resolved by the use of the senses.

However bewildering it may be, we are forced to think that the metaphysical view that there are no material things and the commonsense truism that
there are cannot be evaluated by sense observation. The senses can play no
role in resolving a *philosophical* disagreement regarding the existence of a
category of objects. In an *ordinary*, non-philosophical case in which the
existence of something in a certain region is in dispute, for example, the Loch
Ness Monster, it is possible, *in principle*, even though not in fact, to come to a
decision one way or another to the satisfaction of the parties to the dispute:
we can describe an encounter with the Loch Ness Monster and also describe a
series of investigations which conclusively establish its nonexistence. Furthermore, in the case of an *ordinary* disagreement over whether something is
really taking place or only appears to be taking place, e.g., whether a neon
arrow is really in motion, it is again in principle possible to determine by
sense tests what the actual state of affairs is. The *philosophical* debate regarding the status of the testimony of the senses is entirely different in character.
For there is no describing a situation which would show that the use of the
senses was justified any more than there is a way of describing one which
would show that their use was not justified. The judge at the trial of metaphysics versus common sense cannot be sense-experience.

This conclusion throws Moore's well-known translations into the concrete
into a different light from the one we have been used to. According to Moore
the philosophical view that matter does not exist has such translations into the
concrete as 'The earth does not exist' and 'I do not have a body'. It will be
apparent that if these concrete propositions are such that sense-experience is
relevant to determining their truth-values, then in his *philosophical* debate
with metaphysicians like Bradley, Moore is effectively stalemated. And he is
stalemated only because the concrete commonsense propositions to the effect
that I have a body and the earth exists are such that sense-experience is not
relevant to the determination of their truth-values. These statements look to
be empirical, and they appear to be referring to objects of sense-perception.
In the philosophical debate something mysterious happens to them. The
words, 'The earth exists', come from ordinary language, but in the philosophical debate they seem to have become semantically transmuted and to
have lost their ordinary use. No one is a stranger to the fact that the Socratic
corner of the Agora is in some inexplicable way out of touch with the rest of
the market place, in the way in which a dream is out of touch with the

realities of everyday life. This fact stands out, but what we lack is an explanation of it. Not to pursue this, however, the important thing, which cannot be highlighted too much, is that in philosophy sense-perception plays no role in the investigation of a theory or of what appears to be a concrete factual statement which goes against the theory. Parmenides' words come to mind at this point. Having urged us not to take the false path of sense-experience, he enjoins us to bring to the philosophical investigation of reality 'the test of reason'.

The test of reason can be nothing other than the analysis of concepts, or to use Moore's well-known way of putting it, the analysis of the meanings of words. The implication is that Moore's only possible procedure for evaluating philosophical propositions is analysis, which is a wholly *a priori* procedure. On this account of the mode of investigation open to Moore, his translations into the concrete of the proposition that matter does not exist turn out to be propositions for whose evaluation only analytical techniques would seem to be relevant. This way of looking at Moore's translations is in agreement with the fact that the metaphysical view that matter does not exist is a conclusion from an argument which professes to show the concept *matter* to be *self-contradictory*. An argument that establishes or professes to establish a contradiction in a concept or a proposition is a piece of analysis; and, in general, a proposition whose truth-value is determined by analysis is *a priori*.

It will be plain, without arguing the matter, that an *a priori* claim cannot have non-*a priori* consequences, and thus that the proposition that matter does not exist (which carries with it a factual, empirical air) is not logically different from the proposition that the concept *matter* is self-contradictory. A little reflection makes it clear also that Moore's translations into the concrete of the philosophical view that matter is self-contradictory cannot be the empirical propositions they appear to be. Perhaps a brief explanation will be useful here. Sometimes, as in the case of G.H. von Wright,[5] a philosopher will assert that necessary p entails that p is true as a matter of fact, and that impossible p entails that p is false as a matter of fact: $\sim \Diamond \sim p \rightarrow p$ and $\sim \Diamond p \rightarrow \sim p$. If, however, being false *as a matter of fact*, in distinction to being false by logical necessity, implies the logical possibility of the proposition having a truth-value other than the one it has, then a factual falsehood cannot be entailed by a logical impossibility. The same holds for a logical truth and a factual truth. Wittgenstein stated that '. . . *only* tautologies *follow* from a tautology' (*Tractatus* 6.126). The more general statement, that the consequences of *a priori* truths are themselves *a priori*, also holds. To suppose otherwise, as von Wright seems to, is to adopt a position which implies a contradiction: namely, the contradiction that a proposition can be both possibly false and necessarily true. It becomes immediately evident that Moore's concrete translations cannot be the factual consequences they seem to be.

We arrive at a similar result if we consider what Moore said in *Some Main Problems* about Berkeley's theory regarding material objects. Berkeley and many present-day philosophers take the theory as being about the *nature* of things, as disclosing or professing to disclose the 'inner nature of reality', to use Quine's phrase. But with Samuel Johnson, and students who have not yet been initiated into the ways of philosophy, Moore maintained that Berkeley's view implies that there are no material objects, which is to say, concretely, that things like stones, human bodies, and shoes do not exist. Moore wrote: '[Berkeley] says that [he] is not denying the existence of matter, but only explaining what matter is. But he has been commonly held to have denied the existence a matter, and, I think, quite rightly . . . he denies the existence of any material objects, in the sense in which Common Sense asserts their existence. This is the way in which he contradicts Common Sense.'[6]

One of Moore's reasons for his conclusion is given in the following words: 'Berkeley held that. . .these Appearances do not exist except at the moment when we see them; and anything of which this is true can certainly not properly be said to be a material object: what we mean to assert, when we assert that the existence of material objects, is certainly the existence of something which continues to exist even when we are *not* conscious of it.'[7] Moore's way of putting the matter, with its muted reference to the meanings of terms in everyday use, suggests the idea that Berkeley's view is not an empirical hypothesis. Rather, it suggests that the view is the outcome of an analysis of the concept 'material object' and thus that it advances an entailment-claim: the claim that being a sensible appearance *entails* being perceived. Moore's own statement about 'what we mean to assert' when we declare the existence of a material thing is also an entailment-claim, namely, the claim that being a material thing entails being such that its existence does not depend on its being perceived. The implication of this is that a material object cannot be identified with *any* set of sensible appearances. On Moore's own account the implication of Berkeley's position that to be is either to be perceived or to be a perceiver, which is restatable as the entailment-claim that to exist entails being either a sensible appearance or a mind, is that it is *logically impossible* for matter to exist. Concretely, it implies that it is logically impossible for there to be stones, shoes, or human bodies. It is easy to see that a similar conclusion can be obtained with regard to Leibniz's view that a material thing is a colony of monads.

It will be clear, in general, that a translation into the concrete of an *a priori* statement will itself be an *a priori* statement. For example, the proposition that 4 is not a prime number, which is implied by the *a priori* statement that no even number greater than 2 is a prime number, is itself *a priori*. It will be equally clear that a translation into the concrete of a putative *a priori* statement, e.g., the statement that the existence of space implies a contradiction, will itself be a putative *a priori* statement. Now, the analysis of a concept or of a set of concepts can in principle discover for us the truth-value of a purported *a priori* statement, which is to say that it can in principle enable us to decide whether a claimed entailment *is* an entailment. Concept analysis is not, however, a technique for establishing the truth-value of an empirical statement. It is not a method for calculating the truth-value of a statement whose actual truth-value is not its only theoretically possible truth-value. It can determine for us whether one empirical proposition is entailed by another empirical proposition or set of propositions, but it is not a procedure for determining which one of its *possible* truth-values an empirical proposition actually has. The von Wright thesis that what is necessarily true is also true as a matter of fact can be seen to go against this maxim about what analysis can do. If a factually true proposition could be a logical consequence of a necessary truth, then since the truth-value of the necessary proposition is, in principle, analytically determinable, analysis could, in principle, be used to calculate the truth-value of at least some factual statements, namely, those that are consequences of *a priori* ones.

Linked with the idea that an *a priori* proposition has factual consequences is the further idea that some *a priori* propositions are about the world. Some philosophers unquestionably labor under this notion, and probably all do. Kant's view with regard to a special class of *a priori* truths, namely those the predicates of which are neither identical with nor a conjunctive part of their subjects, is that they are about phenomena; to put it in his way, they are about how the sensible manifold works. As is known, some philosophers, in disagreement with the view that the sole function of analytic propositions is to clarify concepts, even have the idea that analytic propositions are about the world, about how things must be as well as how we are constrained to think about them. Leibniz, for instance, maintained that an identical proposition is a truth that holds for all possible worlds, which carries with it the idea that it is a truth about the world that exists, and about alternative worlds as well. To inject a personal note, when I visited Mrs. G.E. Moore in the summer of 1970, I noticed a copy of my *Structure of Metaphysics* among the books in Moore's former study, which turned out to be annotated by Moore. In one place in that book I make the statement that 'a proposition that is true or false by logic can say nothing about the world', and in another place I state that if theories are about phenomena they cannot be *a priori*. In each instance Moore wrote in the margin the question, 'Why not?' I shall shortly give reasons for making these assertions. But first it needs to be said again that observation is not available to philosophers as a procedure for investigating reality, for determining what there is and what there is not. If analysis is the only instrument of investigation available to a philosopher, then if it can be shown that an *a priori* statement yields no information about the world, it will follow that Moore's program as laid down in the first few pages of *Some Main Problems* cannot be carried out.

When Moore asked why a proposition that is true or false by logical necessity is empty of ontological content, he knew of one philosophical theory about the nature of logical necessitation which implies that an *a priori* truth gives no information about things. Conventionalism maintains that a necessary truth conveys only information about usage, that it 'records our determination' to use words in certain ways. The consequence of this is that a sentence which expresses a necessary proposition does not use words to make a statement about things. Apparently, however, the conventionalist's implied claim that a necessarily true proposition has no ontological import was not the answer that Moore was seeking. It is plausible to think that he had the idea that there might be a different reason for holding this. And, indeed, there is. Consider first a tautological statement of the form $p \vee \sim p$, for example, the statement 'It is either raining or not raining'. Wittgenstein remarked in the *Tractatus* (4.461) that we know nothing about the weather when we know it is either raining or not raining. We might add that in order to learn that it is either raining or not raining it is not necessary to go to the window and look out. Observation plays no role whatever in determining its truth-value, for the reason that it makes no declaration about the weather. If, for example, we read in a weather prediction that tomorrow it will either rain or not rain, we should certainly say that we had been told *nothing* about what the weather tomorrow will be. We might say of a meteorologist who gave only 'tautological predictions' that he is never wrong, but we should stop reading his reports. At this kind of reporting those who are innocent of meteorology are as authoritative as the professional.

The reason why a statement of the form $p \lor \sim p$ is barren of information about states of affairs, past, present, or future, is that it is logically compatible with every conceivable state of affairs. Consider the truth-tables for $\sim p \supset q$ and for $\sim p \supset q \cdot \lor \cdot \sim p \cdot \sim q$, which are usually given as follows:

p	q	$\sim p \supset q,$	$\sim p \supset q \cdot \lor \cdot \sim p \cdot \sim q$
T	T	T	T
T	F	T	T
F	T	T	T
F	F	F	T

It is evident that the first three truth-conditions are consistent with the truth of $\sim p \supset q$ and that the fourth is inconsistent with it. We thus have the following consistency-inconsistency table (0 = consistent with, $\emptyset$ = inconsistent with):

$$
\begin{array}{cc}
p & q \\
\hline
T & T \\
T & F \\
F & T
\end{array} \Bigg\} \quad 0 \quad \sim p \supset q
$$

$$
\begin{array}{cc}
F & F
\end{array} \quad\;\; \emptyset \quad \sim p \supset q
$$

The truth-table for the tautology yields the following consistency table:

$$
\begin{array}{cc}
p & q \\
\hline
T & T \\
T & F \\
F & T \\
F & F
\end{array} \Bigg\} \quad 0 \quad \sim p \supset q \cdot \lor \cdot \sim p \cdot \sim q
$$

We can see that $\sim p \supset q$ is not true independently of its truth-conditions, as one of its truth-conditions makes it false. The formula $\sim p \supset q \cdot \text{ v } \cdot \sim p \cdot \sim q$, however, it true *independently* of its truth-conditions. In fact the set of truth-conditions of the tautology are just the tautology itself in an expanded version. Leibniz' dictum that an identical proposition is true for all possible worlds is merely an arresting way of saying that it is *consistent with* all possible worlds, which implies not only that no world would make it false but also that none makes it true. It is not a quibble to say that a proposition is *made* true by the occurrence of a state of affairs only if it could be false, i.e., be open to theoretical falsification. A proposition which cannot be made false by any possible state of affairs, because it is consistent with all of them, is not *made* true by any of them. To say that a proposition is *true* for all possible worlds is just another way of saying that it is *consistent with* every state of affairs, that no possible state of affairs would upset it. This in turn is just another way of saying that it is true *a priori*. But the Leibnizian and also the standard truth-table way of exhibiting tautologies represents them as irrefutable, absolutely secure *truths about things*. F.P. Ramsey characterized tautologies as 'not real propositions, but degenerate cases',[8] which seems to recognize in an oblique way the underlying logical difference between tautologies and contingent statements. Following Wittgenstein's *Tractatus* he stated that a tautology 'really asserts nothing whatever; it leaves you no wiser than it found you'.[9] We might say that it asserts nothing whatever about *things*; it leaves you no wiser about *things* than it found you. In sum, a tautological proposition cannot be upset by what there is and so says nothing about what there is. Some philosophers go on to the further thesis that it says *nothing*, but that is another matter, which cannot be gone into here.

The general point which emerges is that a proposition whose truth-value is invariant under all theoretical conditions makes no declaration about the world. This holds not only for tautologies but for every proposition whose actual truth-value is its only logically possible truth-value. Kant maintained that there is a special class of propositions which are characterized by 'inner necessity' but whose predicates lie wholly outside of their subjects, i.e., they are synthetic but possess their truth-value by logical necessity. In his view, unlike a sentence which expresses an analytic proposition, a sentence which expresses a synthetic *a priori* proposition conveys information that augments our knowledge of what its subject term refers to. If a proposition is *a priori* however, regardless of whether it is analytic, it is in Leibniz's way of speaking consistent with all possible states of affairs, and is as empty of ontological information as a tautology. For it is true no matter what the world is like and no matter what there is or is not, and thus makes no claim about what the world is like or about what there is. Not only is no necessary proposition ampliative about things, it makes no declaration whatever about them.

As is well known, Kant took the proposition that everything which happens has a cause to be both synthetic and *a priori*. Like the analytic proposition that every effect is a caused occurrence, it has, in his view, 'the character of necessity, and therefore is completely *a priori*' but unlike the analytic proposition it is ampliative and gives us secure knowledge of the way the world operates. However, *if* the truth-value of the proposition that every occurrence has a cause is decidable 'on the basis of mere concepts', then, like the analytic proposition that every change which is an effect has a cause, it gives no more information about the conditions under which changes occur than does the analytic proposition. For since it possesses its truth-value by 'inner necessity', it cannot be made false by any theoretically possible world and, hence, like a tautology, is consistent with all possible worlds. It thus is empty of ontological information, true or false, about any possible world.

The point that the proposition, 'Every change has a cause', viewed as necessarily true, equally with the analytic proposition, 'Every effect has a cause', is empty of information about how change occurs can perhaps be brought out more perspicuously in the following way. It is easily seen that if the proposition expressed by the *sentence* 'Every change has a cause', or 'Every change is a caused change', were true by logical necessity, the term 'caused' would not function in the sentence to distinguish between occurrences of change. The terms 'caused change' and 'change' would not differ descriptively *when used to refer to occurrences*: to say that x is a caused change would convey no more information about x than to say that it is a change. Thus, the sentence 'Every change has a cause' would say nothing different about what the term 'change' applies to than is said by the tautological sentence 'Every change is a change'. The claim about occurrences of change made by the putative synthetic *a priori* proposition would be no different from the claim about the occurrence of change made by the analytic proposition, and is as empty of factual import as it is.

On the other hand, if the function of the sentence in the language is to express a synthetic proposition the truth-value of which is not decidable 'on the basis of mere concepts', then the term 'caused' will serve to distinguish among changes and the factual information conveyed by saying that x is a caused change will not be identical with the information conveyed by saying that x is a change. In that case the proposition would not be one whose truth-value or probable truth-value can be evaluated by techniques available to philosophers. In his *Notebooks, 1914–16* Wittgenstein stated that all of his philosophical problems revolved around the question: 'Is there an order in the world *a priori*, and if so what does it consist in?' We might say that if he had discovered the order he was seeking by the examination of concepts alone, he would have come into possession of a proposition which says nothing about the world. But if in his quest he had discovered a proposition that makes a claim about the order of the world, he would have had to seek elsewhere than in philosophy for techniques relevant to establishing its truth-value.

THE FALSIFIABILITY THESIS

Professor J.W.N. Watkins takes the position that it is possible for a proposition to have a truth-value and be neither empirical nor *a priori*: '. . .a sentence which expresses neither an empirical proposition (in approximately Lazerowitz's sense) nor a logical truth, may nevertheless express a truth-valued proposition.'[10] In his view, a pure existential statement, such as 'There exists a metal which does not expand when heated', could in principle be shown by investigation to be true but is in principle unfalsifiable. It thus declares the existence of a thing, i.e., it makes a factual claim, but is nonempirical because it is unfalsifiable in an 'infinite space-time region'.

What the connection is between the impossibility of *showing* a proposition to be false by a series of observations or by an examination of cases, and the impossibility or possibility of its being false needs to be looked into further. It can easily be seen that if the logical impossibility of showing a proposition (which could be true) to be false *entailed* the logical impossibility of the proposition's being false, then the proposition would be *a priori* true. Its being possibly true would in that case be equivalent to its being necessarily true.[11] One idea which Watkins appears to have is that a proposition of the form $(\exists x)fx$ could be false but cannot, in principle, be shown by observation or by an examination of cases to be false. Taken at face-value, as making an *a priori* claim, this idea would seem to imply that we could know what it would be like for a proposition to be false without knowing what it would be like to *know* that it is false. ·

Not to pursue this for the moment, however, the point which needs to be made here is that a proposition *counts* as being empirical if its actual truth-value is not its only theoretically possible truth-value, as against an *a priori* proposition whose actual truth-value is its only theoretically possible one. A proposition of the first type is one such that, in entertaining it, we know what it would be like for it to be false; that is, we can conceive a state of affairs which would make it false. This is not so in the case of an *a priori* proposition: if it is false, there is no conceiving a circumstance which would make it true, and if it is true there is no conceiving a circumstance which would make it false.

Watkins maintains that an 'uncircumscribed' existential proposition, e.g., the mathematical proposition that there are seven consecutive 7's in the decimal expansion of π, is such that no examination of intervals or 'empirical findings' can show it to be false. This feature, or apparent feature, of pure (non-mathematical[12]) existential statements is, in his opinion, a sufficient reason for counting them as *not empirical*, while allowing that they have factual content. Nevertheless, they are propositions which could, conceivably, be false: we know what it would be like for there not to be a metal which fails to expand when heated, or for there not to be substances which fail to heat up when in friction. Thus, Watkins' view would seem to come to the claim that some propositions cannot be falsified, in the sense of being *shown* false by observation, but could be falsified in the sense of being *made* false by a conceivable state of affairs.

Instead of saying with Watkins that non-*a priori* pure existentials are 'unempirical' propositions which lie in the 'no man's land between testability and analyticity',[13] I should say that they are empirical propositions which some philosophers hold cannot be upset by any finite number of observations. Looked on as making a claim about a putative feature of uncircumscribed empirical statements of the form $(\exists x)fx$, the view that they are, unlike their circumscribed counterparts, theoretically unfalsifiable implies that it is logically impossible to know that they are false. The linguistic proposition about English usage corresponding to this conclusion is that expressions of the form 'knows that $\sim(\exists x)fx$' are literally senseless.

Taking the linguistic proposition for the fact-claiming proposition it appears to be, it implies that the connected expressions of the form 'believes that $\sim(\exists x)fx$' are also senseless. For in their normal use 'know' and 'believe' are so related semantically that if it makes no literal sense to speak of knowing that p, it also makes no literal sense to speak of believing that p. It is hard to think that any philosopher believes, whether unwittingly or not, that it makes no sense to say, for example, 'Jones believes that there is no centaur anywhere in the universe'. And this, without going into the matter, is a reason for thinking that the unfalsifiability thesis is not the fact-claiming proposition it appears to be. [For an enlightening explanation of a comparable philosophical position see John Wisdom's 'Philosophical Perplexity', *Proceedings of the Aristotelian Society* **16**, 1936].

NOTES

[1] Pp. 1–2.
[2] 'Philosophical Progress in Language Theory', *Metaphilosophy* **1**, p. 2.
[3] 'A Defence of Common Sense', *Philosophical Papers*, p. 33.
[4] Ibid., p. 41.
[5] 'That which is necessarily the case is also as a matter of fact the case.' 'Deontic Logic and the Theory of Conditions', *Crítica* **2**, p. 3.
[6] *Some Main Problems of Philosophy*, p. 21.
[7] Ibid.
[8] *The Foundations of Mathematics*, p. 10.
[9] Ibid.
[10] 'Word Magic and the Trivialization of Philosophy', *Ratio* **7**, p. 214.
[11] It is difficult to suppose Watkins to be holding that some propositions are such that they could be, but need not be, true, and yet are by their very nature incapable of being false, in other words, that they possess the truth-value truth *contingently*, but in failing to be true nevertheless are not false. One implication of this position, which I do not attribute to him, would seem to be that some propositions are both possibly not true and necessarily not false.
[12] The mathematical statement about seven consecutive 7's in the expansion of π and the statement about a metal which does not expand when heated require separate treatment. The nonoccurrence of seven 7's in an interval in the expansion implies the logical impossibility of their occurrence in that interval, whereas the nonexistence of a metal which fails to expand when heated at a given time or place does not eliminate the logical possibility of the metal existing at that time or place. Counting the primes in a segment of the natural number series, for example, and counting the people in a room are usually taken to be comparable processes. To use an expression of George Gamow (*One Two*

Three...Infinity, Viking Press), we learn in each case how many there are 'empirically by counting'. But this way of describing the two kinds of cases carries with it the misleading idea that the statements, 'There are eleven people in the room' and 'There are eleven prime numbers between 2 and 43', are on the same logical footing.

[13] 'Confirmable and Influential Metaphysics', *Mind* 67, p. 363.

CHAPTER VI

MOORE AND LINGUISTIC PHILOSOPHY

Academic philosophy has undergone a striking metamorphosis in recent years, one so radical that a philosphical Rip van Winkel would indeed have to rub his eyes to recognize his subject in it. The attention of philosophers has become more and more concentrated on language; and linguistic considerations, which were once introduced incidentally, for the sake of clarifying a question or an argument, now occupy a central place in the doing of philosophy. It may be that the inner substance of philosophy has remained unchanged and that what philosophers seek now is what philosophers have always sought. But there can be no doubt that the approach to problems has, apparently, gone through a change which is as strange and disconcerting to some able philosophers as it is exciting and promising to others. In the *Phaedo* Socrates says that he tried to determine 'the truth of existence' by recourse to concepts: this is the traditional image of the philosopher. And Moore's distinction between knowing the meaning of a common word and knowing the analysis[1] of its meaning fitted in with the picture of the philosopher looking more deeply into concepts than does the ordinary man and discovering basic facts about things. Wittgenstein, it will be remembered, told us not to ask for the meaning of a word but rather to ask for its use; and in line with this recommendation many present day philosophers examine linguistic usage with the traditional aim, apparently, of establishing facts of ontology.

I

Wittgenstein, who even more than Moore, has directed the attention of philosophers to language, declared that philosophers reject a form of words while imagining themselves to be upsetting a proposition about things, and that what they need in order to rid themselves of their fantasy is a clear understanding of the workings of language. Some philosophers, however, seem to have the idea that the study of language can bring to light basic features of reality, that facts about things can be inferred from facts of linguistic usage. Professor C.D. Broad, who identifies Moore's famous essay, 'A Defence of Common Sense', as an important source of this idea, writes that it

... has led many able men, who might have contributed to solving the real problems of the philosophy of sense-perception, to waste time and labour and ingenuity in semi-linguistic studies of the usages of ordinary speech in the language with which they happen to be familiar. To imagine that a careful study of the usages, the implications, the suggestions, and the *nuances* of the ordinary speech of contemporary Englishmen could be a substitute for, or a valuable contribution towards, the solution of the philosophical problems of sense-perception, seems to me one of the strangest delusions that has ever flourished in academic circles.[2]

Bertrand Russell, who also cannot accept the present day linguistic approach to philosophy, has characterized it as consisting in 'reasoning from the actual use of words to answers to philosophical problems, or from a conflict in actual uses to the falsehood of a philosophical theory'.[3]

A.J. Ayer, in one place at least, provides us with an illustration of the notion that analytical lexicography is capable of yielding ontological information. He tells us that 'what we obtain by introducing the term "sense-datum" is a means of referring to appearances without prejudging the questions what it is, if anything, that they are appearances *of*, and what it is, if anything, that they are appearances to'.[4] This introduction of a term into the language of perception seems aseptic enough; its use is to refer to appearances, pure and simple. i.e., without linking the term to a philosophical theory, and it merely substitutes for expressions which are already in the language. But the point of introducing the term 'sense datum' is not merely to extend philosophical patronage to a word:

The idea is that it helps you to learn something about the nature of physical objects, not in the way that doing science does, but that you come to understand better what is meant by propositions about physical objects, what these propositions amount to, what their 'cash value' is, by restating them in terms of sense-data. That is, the fact that you *can* restate them in this way, if you can, tells you something important about them.[5]

It is not easy to see how what a philosopher wishes to learn about the *nature* of physical objects differs from what science might teach him; but it seems clear, at any rate, that Ayer has the idea that the analysis of linguistic usage, aided by the introduction of the term 'sense-datum', can yield information about the nature of such things as stones and sheets of paper.

Norman Malcolm provides us with an example of a philosopher who 'refutes' a philosophical view by recourse to linguistic usage. In this opinion, H.A. Prichard's philosophical view that we do not really see physical objects is upset by the fact that such expressions as 'sees a sheet of paper' have a use in the language. 'Prichard and others', he writes,

> must admit that we use such sentences as 'See my finger', 'Now you see the dog', 'Now you don't see him', every day of our lives; and furthermore that we are taught to use such sentences and teach their use to others. We are taught and do teach that the correct way to speak, in certain circumstances, is to say 'I see the dog', and in other circumstances to say 'I don't see him now', and in still other circumstances to say 'I think I see him', and so on. Undoubtedly Prichard used such forms of speech every day (and taught them to his children, if he had any) and would have acknowledged in various ways in practical life that they are correct forms of speech. His philosophical position, however, stands in opposition to this obvious fact.[6]

The correct inference to draw, it would seem, is that Prichard's philosophical view is false, and that what shows it to be false is a matter of linguistic usage.

These two examples are sufficient to illustrate Broad's reproach that some philosophers labour under the notion that by recourse to language alone it is possible to solve some, or perhaps all, philosophical problems. One philosopher thinks, or seems to think, that the scrutiny of linguistic usage will reveal facts about the nature of things, another that it is sufficient to refute a philosophical theory about our perception of things; and still a third philosopher, it might be added, professed to demonstrate the existence of free will from the linguistic fact that the term 'free will' has a correct use. To sum up, the idea at work behind linguistic philosophy is that linked with lexicography, which records the usages of common speech, there is something which might be called ontological lexicography, the study of which enables us to determine the truth-values of philosophical theories and to obtain knowledge of what exists, and of what does not exist. Broad characterizes this idea, specifically in connection with the philosophical problem of perception but undoubtedly quite in general, as a delusion.

Taken at face value, it is an odd idea, one which we might not be surprised to find alive in people who have yet to emerge from the stage of magic thinking, but which we should be astonished to find active in the professional work of sophisticated thinkers. No philosopher, regardless of his special persuasion, is so unrealistic in the usual conduct of life as to believe that any sort of study of language will by itself reveal to him that amber has electrical properties, that pelicans exist, and that phlogiston does not exist. If, as appears at first glance to be the case, linguistic philosophers believe that ontology can be learned from the study of linguistic usage, then indeed they suffer from a bizarre delusion, but one, it is hardly necessary to point out, which is confined to philosophy. It is a professional delusion, and must be bound up in some way with the nature of philosophy itself. It would seem that we have to think either that a considerable number of outstanding philosophers labour under a curious idea, one which is wholly inappropriate to their subject, or that the idea is in some way appropriate and that the nature of the subject has not been clearly perceived. We may be less inclined to reject the second alternative if we do not push out of our minds the strange fact that in its long history philosophy has not been able to produce a single uncontested result.

Broad lays the present unwelcome development in philosophy at Moore's door; and there are many things in the 'Defence of Common Sense', undoubtedly the most sober treatise in the history of a discipline not known for its sobriety, which point towards linguistic philosophy. Indeed, it is not too much to say that after Moore the only direction in which philosophy could go was towards linguistic analysis. And there are many things, not only in his 'Defence of Common Sense', but in a number of other papers, which propelled philosophy in the direction it did take. To mention one of Moore's observations, his pointing out the shocking fact that philosophers have been able to hold sincerely philosophical positions inconsistent with what they knew to be true[7] could not fail to have had great impact on the thinking of philosophers, all the greater, perhaps, for doing its work in the underworld of the mind. If the philosophical statement that Reality is an undifferentiated unity is inconsistent, for example, with the proposition that there are fifteen students, twelve men and three women, in the lecture room, the inconsistency is *conspicuous*; and if the philosophical statement that no one can really know that another person feels pain is inconsistent with the proposition that I *know* that Jones feels pain (because he winces whenever I jab him with a pin), the inconsistency is *conspicuous*. It is not a trade secret that philosophers are not made to give up their everyday statements by their professional views, nor their professional views by their everyday statements. The strange situation to which Moore's Paradox has directed the attention of philosophers demands an explanation, and continues to do so even when pushed out of the forefront of one's thinking. The attempt to explain it can, in the end, lead us in one direction only.

Many philosophers have, as was to be expected, taken refuge in what might be called intellectual solipsism. Like Galileo's colleagues who would not look through his telescope, they go on with their work as if Moore had never stated his paradox. If forced to give *some* explanation of the situation, they reassure themselves with the Berkeleian formula: speak with the vulgar, but think with the learned. But there is no serious question that they not only speak with the vulgar but that they also *think* with the vulgar. Their using the speech of everyday life is not a hypocritical concession, intended to spare them the rude comments of the gross. Their beliefs are the beliefs of the ordinary commonsense man, and in consequence of Moore's paradox their taking a stand with philosophy against whole classes of everyday statements has come to acquire an air of unreality. The impression created is that a *game* is being played.

Another possible explanation, one which seems to be implied by Moore's Paradox, is that philosophers really believe their philosophical statements. But this is to imply that philosophers actually suffer from strange delusions, for example, that they do not really know that others feel pain or that there is a number of students in a room. The picture of the philosopher which is conjured up is that of a person who simultaneously lives in two worlds, the ordinary world of everyday experience and everyday talk and an exotic world of odd convictions. But although this is a possibility, it is certainly not a realistic possibility. There is no observational evidence that the picture applies to the philosopher, and it offers no satisfactory explanation of the situation to which Moore has directed our attention.

One path remains open. This is to suppose that there is no inconsistency between the philosophical views which appear to go against everyday propositions and the everyday propositions themselves. But to suppose that there is no inconsistency between what a person says on a given occasion who uses the sentence 'I know that he is in pain' and what the philosopher expresses by the sentence 'No one really knows whether anyone else feels pain' is to imply that the two sentences are not about the same subject matter, that they do not refer to the same sort of thing. What could not fail to suggest itself, independently of various things that Moore said, is that the philosophical sentence, despite the nonverbal mode of speech in which it is framed, makes a verbal claim, to the effect that the expression 'knows that someone else feels pain' has no correct application.

A number of observations Moore made prodded philosophers in the direction of a linguistic interpretation. In 'Some Judgments of Perception' he wrote: 'This, after all, you know, really is a finger: there is no doubt about it: I know it, and you all know it.'[8] The strong impression these words make on one is that a pretence is being exposed and that philosophers are being cajoled out of it. But the pretence obviously is not like that of the youngest son in the fairy tale who pretends stupidity. One idea which is bound to suggest itself is that, whether consciously or not, Moore is trying to free philosophers from a wrong way of looking at their own utterances, from a misunderstanding of their views, and that he is trying to straighten them out about facts of ordinary language.

In his 'Defence of Common Sense' he remarked:

In what I have just said, I have assumed that there is some meaning which is *the* ordinary or popular meaning of such expressions as 'The earth has existed for many years past'. And this, I am afraid, is an assumption which some philosophers are capable of disputing. They seem to think that the question 'Do you believe that the earth has existed for many years past?' is not a plain question, such as should be met either by a plain 'Yes' or 'No', or by a plain 'I can't make up my mind', but is the sort of question which can be properly met by: 'It all depends on what you mean by 'the earth' and 'exists' and 'years'. if you mean so and so, and so and so, and so and so, then I do; but if you mean so and so, and so and so, and so and so, or so and so, and so and so, and so and so, or so and so, and so and so, and so and so, then I don't, or at least I think it is extremely doubtful.' It seems to me that such a view is as profoundly mistaken as any view can be. Such an expression as 'The earth has existed for many years past' is the very type of unambiguous expression, the meaning of which we all understand.[9]

This, as a moment's sober reflection shows, is not a caricature of the philosopher, nor of philosophical discussions. When an important and greatly respected philosopher holds up the mirror to his colleagues the image that looks back at them must make an impression which cannot simply be shrugged off. It leaves an active residue in the mind.

One more of Moore's remarks should be recalled. In 'Some Judgments of Perception' he wrote:

Some people may no doubt think that it is very unphilosophical in me to say that we *ever* can perceive such things as [doors and fingers]. But it seems to me that we do, in ordinary life, constantly talk of *seeing* such things, and that, when we do so, we are neither using language incorrectly, nor making a mistake about the facts — supposing something to occur which never does in fact occur. The truth seems to me to be that we are using the term 'perceive' in a way which is both perfectly correct and expresses a kind of thing which constantly does occur, only that some philosophers have not recognized that this is a correct usage of the term and have not been able to define it.[10]

Joined to his remark two pages later that 'The questions whether we ever do know such things as these, and whether there are any material things, seem to me therefore, to be questions which *there is no need to take seriously*',[11] the point is clear. What needs to be taken seriously is the *failure* by philosophers whose views go against everyday statements to 'recognize' cases of correct usage. Wittgenstein said that 'philosophical problems. . . are, of course, not empirical problems; they are solved, rather, by looking into the workings of our language',[12] and that what philosophers need is to get a 'clear view of our use of words'.[13] There can be no doubt that Wittgenstein had his most important roots in Moore, whose defence of common sense began to emerge as a defence of the language of common sense, i.e., the language in everyday use.

There has been disagreement in recent years over whether the linguistic interpretation of Moore's defence of common sense is correct. I might say that in the course of a philosophical discussion in Cambridge Moore told me that he accepted the interpretation Norman Malcolm placed on his defence in 'Moore and Ordinary Language'. To introduce a personal note, he rejected forcefully, even with some heat, my own account of what his defence came to, but thought that the account which made him out to be correcting philosophers' mistaken ideas about usage was correct. Moore's work made many philosophers aware of the linguistic character, or what first appeared to be the linguistic character, of a particular type of philosophical theory, and Wittgenstein went on to throw a linguistic shadow over the whole of philosophy. The idea which began to take its place among the standard ideas as to *what* philosophy is, to put it somewhat metaphorically, is that language is the stuff philosophy is made of.

What Moore's Paradox could not fail to suggest is the idea that philosophical theories which seem to be inconsistent with ordinary assertions are *not* inconsistent with them, but instead go against the *language* used to give them utterance. This is the idea which emerged about the nature of paradoxical theories, those which appear to flout common sense; and this idea can easily be seen to apply also to more sober and more scientific sounding theories which seem antithetical to large classes of ordinary statements. To illustrate, take the philosophical question regarding what is implied by the physical fact that light takes time to travel from an object to its perceiver. Does this physical fact imply that we never see a material thing, e.g., a book that is being read, at any of the times it seems to us that we are seeing it? Light waves from the nearest star take four years to reach the earth, which implies, according to some philosophers, that a star gazer, to whom it seems that he is 'seeing' the star, is not in fact seeing it. At best, it is argued, he can be said to be perceiving what was in the past, not the star as it is at the time of his perception, if indeed it is still in existence then. The philosophical outcome is that we do not see remote objects such as stars. But like Zeno's argument which applies to fractions of a distance to be traversed as well as to the entire distance, the consideration which applies to astronomically far objects applies equally to near objects. The consequence, which is drawn by many philosophers, is that at no time do we see the thing it seems to us we are seeing, and thus that we do not see physical things. Some philosophers, under the influence of things Moore has said, would place a linguistic interpretation on this claim, rather than, as is natural, a scientific interpretation, the reasoning being somewhat as follows. A philosopher who holds that we never see things at the times we seem to be seeing them implies that the ordinary use of 'see' as it occurs in such phrases as 'sees a sheet of paper' and 'sees a finger' is incorrect. He cannot, to use Wittgenstein's expression, say *what it would be like* to see things, as against only seeming to see them, and is thus holding, directly or indirectly, that such an everyday phrase as 'sees a finger' has no use to describe anything. The view that we do not really see things does not say anything about what does not or cannot happen physically. It declares the logical impossibility of seeing physical things, and thus implies that such an ordinary expression as 'sees a finger', like 'whole number which is half of 7', has no use in the language.[14]

Once having brought out what he takes to be the linguistic character of the view, once having made explicit the verbal claim behind the nonverbal sentence 'We do not really see things at the times it appears to us we are seeing them', a so-called ordinary language philosopher would bring linguistic evidence against the view, by pointing out that perception words like 'sees' have a correct use which is not confined to their occurrence in phrases like 'seems to see'. It will be clear that a linguistic philosopher who infers that we do really see things from the fact that such an expression as 'sees a sheet of paper' has a correct use is pointing out that it is logically possible to see things. The proposition his linguistic evidence supports is of the same type as the proposition that his linguistic evidence is intended to upset: what it is brought against is the claim that it is logically impossible to see things at the times we would be said to see them; and therefore what it is adduced for is the proposition that it is logically possible to see them at those times. In fact, the inference from usage to the philosophical proposition that we do see things consists of nothing more than giving two reports of the same fact in two different modes of speech: the verbal report that expressions like 'sees a sheet of paper' has a correct use and the indirect nonverbal report of the same fact in the idiom of logical possibility.

One important outcome of Moore's Paradox is the position which construes philosophical views opposed to common sense as attacks on everyday language. This is certainly one of Wittgenstein's constructions, as it is also of some philosophers after Wittgenstein. Where Moore says that philosophical doubts about the existence of things, etc., need not be taken seriously, Wittgenstein tells us that philosophical attacks on common sense are not to be countered by 'restating the views of common sense'.[15] The relevant and correct way to deal with them is to straighten out 'misunderstandings concerning the use of words'.[16] The implication, quite apart from other of Wittgenstein's remarks, is that they are, in substance if not in outward appearance, misdescriptions of ordinary language, to be corrected by recourse to the actual use of terminology.[17] This explanation of the paradoxical utterances helps us understand how a philosopher can say the odd things he says without embarrassment and without making others uneasy about him, but it raises a question on which it throws no light. It tells us why a philosopher who attacks common sense is not corrected by the facts and how he can make his pronouncements in the presence of the facts. But it does not help us understand what makes it possible, supposing it to happen, for a philosopher persistently to misdescribe usage in the presence of actual usage. It leaves unexplained why he does not, to use Moore's word, *recognize* actual usage, which he does not give up, nor why, instead of accepting linguistic correction, he takes refuge in such questions as 'What is ordinary language?' and 'Does the fact that a philosophical term occurs in a standard dictionary make it part of ordinary language?' Nevertheless, this linguistic way of looking at many philosophical theories, the sober alike with the unsober, improves our understanding of the troubling question as to why they can remain in unending debate.

It should be noted that there is a further kind of philosophical position which is not open to the interpretation of describing, truly or falsely, ordinary language (or to the interpretation of misusing language). This instead seems open to the charge that it misconceives the causal powers of language. W.V. Quine, for example, gives the impression of thinking that the contents of the world can be changed by changing linguistic categories.[18] Occam laid down the principle that entities should not be multiplied unnecessarily, which, apart from the Platonic theory it was directed against, would be taken to refer only to the *postulation* of entities. In the view of some philosophers, so it would seem at any rate, by manoeuvring with language ontology can be enlarged or decreased. For example, by widening the category of singular terms, entities are introduced into the world, are brought into existence; and by narrowing the category of singular terms, by not counting indicative sentences as names, for instance, certain kinds of objects are barred from existence. Quine has written: 'When we say that some dogs are black, $(\exists x)$ (x is a dog.x is black), we explicitly admit some black dogs into our universe, but we do not commit ourselves to such abstract entities as dog-kind or the class of black things; hence it is misleading to construe the words 'dog' and 'black' as names of entities.'[19]

We do not, of course, 'admit' black dogs into the universe by *saying* that some dogs are black, any more than an Irishman actually peoples Erin with leprechauns by saying that some leprechauns have beards. Not having the powers of Divinity, saying so-and-so does not bring so and so about, and it is out of the question to suppose that any philosopher might think the contrary, regardless of the suggestions his language carries with it. But in philosophy things are not as they are in ordinary life. There is an atmosphere of bewitchment in it that is not encountered in ordinary work and talk, and language which would be considered inappropriate in connection with an everyday statement, as just a way of speaking, might not be just a way of speaking when used in the expressions of a philosophical position. Thus, when Quine writes that it is important 'to show how the purposes of a segment of mathematics can be met with a reduced ontology' the implication of the words is that ontology can be reduced, i.e., that the number of things there are can be decreased. And the suggestion created by his language has to be taken seriously: the implication of the words is not merely that a decreased ontology can be postulated but that ontology can be decreased by grammatical fiat. How this is to be effected may be gathered from a number of different remarks. The following, in addition to those already cited, should suffice: 'In discussing the theory of meaning I did urge the uselessness of meanings as entities. My thinking was, in effect, that no gain is to be sought in quantifying over such alleged entities as meanings',[20] and 'I prefer not to regard the general terms and statements as names at all'.[21] The idea which comes through is that by 'widening the category of terms that name'[22] entities considered by Quine to be useless are *made* part of ontology, that by grammatically turning 'dog' and 'Friction generates heat' into names or singular terms, abstract entities are created, and that by excluding such expressions from the category of names, entities are barred from existence. The idea appears to be that ontology, the content of the world, is subject to control by grammatical fiat.[23] In one place Ayer speaks of creating a new domain of private entities by a 'stroke of the pen', and from other philosophers the comparable impression is gained that by grammatical gerrymandering, i.e., by manoeuvring with linguistic categories, various kinds of abstract entities are either made part of what is or subtracted from what is. Indeed, the thought which forces itself on one is that unlike such everyday things as shoes and tables, abstract entities are the production of language. To extend Broad's Paradox to philosophers of this general persuasion, we should say that some linguistic philosophers labour under the delusion that by changing language in certain ways reality itself is made to suffer change. It will be clear that the explanation which has been applied to philosophical views falling under Moore's Paradox does not apply to this position.

To come back to the linguistic outcome of Moore's defence of common sense, there can be no doubt that Moore has changed the course of philosophy. His perceptions started a trend that has continued on its own. After Moore it was a natural step to proceed from the notion that a false philosophical theory is one which misdescribes actual usage to the notion that a true philosophical theory is one which correctly describes usage. The whole of philosophy becomes linguistic in character. Thus, Wittgenstein in the Yellow Book speaks of the 'confusion which considers a philosophical problem as though such a problem concerned a fact of the world instead of a matter of expression'.

Many philosophers have viewed the linguistic turn which philosophy has

taken since Moore with the greatest consternation, since it deflates philosophical questions into 'mere questions about words'.[24] Broad, who deplores
the effects of Moore's 'Defence of Common Sense' on the doing of philosophy, is led to think that linguistic philosophers must suffer from the
delusion that from an examination of word usage they can learn facts about
things. By deciphering an ancient piece of writing, which requires learning the
use certain symbols once had, historical facts are learned; but the facts are
learned from the statements into which the symbols enter and not from the
fact that the symbols have a certain use. Anyone who thought that the study
of the use of terms alone will yield information which is more than just
information about a language, that it is capable in addition to yielding information given by statements using the terminology, would be suffering
from a delusion of some sort.[25] And a philosopher who has the idea that the
study of perception-terminology, for example, could lead to an explanation
of what takes place when a person is said to be perceiving a thing would also
be suffering from an intellectual delusion, one comparable to the idea that it
is possible to discover the properties of a substance by studying its name.
Moore's Paradox, which is about traditional philosophers, has led to a way of
doing philosophy which has created a paradox of its own, one which might be
called Broad's Paradox of Linguistic Philosophy. This is that it is a strange
fact that many philosophers actually believe they can solve 'real' philosophical problems by studying the usage of ordinary speech.

Broad's description of linguistic philosophers has two obvious implications.
One is that the real problems of philosophy are problems about the nature
and existence of phenomena of various sorts, the solution of which will
establish 'truths of existence'. The other is that linguistic philosophers want
from philosophy what philosophers have always wanted from it, or have
appeared to want from it. A sceptical note is perhaps not out of place here.
Freud has remarked that what we get from science we cannot get elsewhere,
and it only calls attention to what cannot have escaped anyone to remark
that for information about *things* philosophers along with others go to
science. Not to pursue this, however, Broad is undoubtedly right in thinking
that linguistic philosophers seek what conventional philosophers seek, that
their goals are the same, although their approaches appear to be utterly
different. The words of one important linguistic philosopher are a clear
enough indication of this. J.L. Austin has written: 'When we examine what

we should say when, what words we should use in what situations, we are looking again not *merely* at words (or "meanings", whatever they may be) but also at the realities we use the words to talk about: we are using our sharpened awareness of words to sharpen our awareness of, though not as the final arbiter of, phenomena'.[26] Parenthetically, it is interesting to compare Austin's statement that a sharpened awareness of words is not a 'final arbiter' of phenomena with Ayer's observation that a philosopher who adopts the 'sense-datum' notation learns something about the nature of physical objects but not what a physicist by his work learns about it. Both seem to hint at a kind of recognition of what philosophy cannot give us and thus of what it does give us, what its actual 'domain of discourse' is.

The question which comes up is whether the 'real' problems of philosophy are the kind of problems whose solutions, like those of chemistry, will establish facts about the existence and behaviour of phenomena. Moore's Paradox throws much of traditional philosophy into a strange light, some of the strangeness of which is removed by placing a linguistic interpretation on it. Broad's paradox throws linguistic philosophy into a strange light, on the assumption that the traditional problems of philosophy are not questions about the uses of words. But since Moore and Wittgenstein this assumption can no longer be taken for granted. It cannot be taken for granted any longer that traditional philosophy, behind the fact-stating idiom it employs, is different in content from that of linguistic philosophy: the one may be related to the other as the Colonel's Lady to Judy O'Grady. What looks like a bizarre delusion on the part of sophisticated thinkers may be an impression induced by their use of a double idiom, the verbal idiom of linguistic analysis and the ontological idiom of traditional philosophy. There is no question that the passage from statements about usage to statements about what is logically impossible and what is logically necessary, what 'really' is not and what 'really' is, creates the illusion of inferences being made from words to things: the passage, say, from 'The expression "did it of his own free will" has a correct use' to 'Free will does really exist' would certainly create such an illusion, *if* the inferred statement were an ontological reformulation of the verbal statement. And the work of Moore and Wittgenstein has shown that what seems obviously the case to philosphers, for example, that the philosophical sentence 'Free will exists' states the existence of a kind of action, may not be the case at all. It may well be that no philosopher, regardless of his creed and method, has ever *in fact* sought knowledge of things in philosophy, and that his appearing to do so has been a semantically generated illusion from which Moore's work has, in some measure, helped free us.

II

Moore, at least at times, accepted the interpretation of his defence of common sense as a defence of ordinary language against the attacks on it by metaphysical philosophers. It is perhaps not out of relation to this interpretation that Wittgenstein said, 'What *we* do is to bring words back from their metaphysical to their everyday usage'.[27] Moore's writings made many philosophical views an embarrassment to philosophy – such views as that we don't really know that anyone else has thoughts and feelings and that time is unreal; and placing a linguistic interpretation on them helped remove some of the strangeness that had settled over philosophy. But Broad's complaint that linguistic philosophers labour under the delusion that the scrutiny of verbal usage can yield ontological information carries with it the suggestion that *underneath* no philosopher gives up the idea (which he keeps at a distance from himself) that philosophical theories actually have the factual content they appear on the surface to have. If true, this would help us understand why philosophers who have taken refuge in the linguistic interpretation of the more bizarre utterances fail to see a comparable strangeness in the linguistic interpretations themselves, a strangeness which counts as much against them as the strangeness of the factual interpretations counts against the more natural readings. Construed as descriptions of the actual use of terminology, they are flagrant misdescriptions; and to attribute them to philosophers would, indeed, be nothing less than to suppose them to suffer from some sort of aberration. Moreover, the views have the mystifying feature of never intruding themselves into everyday talk, although they have been announced over and over again for centuries. This cannot be explained on the view that they are *mere* misdescriptions of everyday talk.

The linguistic construction placed on philosophical views which flout, in appearance at least, elementary common sense does offer a possible explanation of how philosophers can say the unbelievable things they seem to be saying. As has already been remarked, this is that they are not actually saying what they appear to be saying: their utterances are verbal, about the actual functioning of expressions in a language. But represented as having verbal content it is no less unbelievable to attribute them to philosophers than it is when they are represented as having factual, nonverbal content. The philosophical utterance, e.g. 'No one can know that anyone else exists' is no less wild when construed as having the verbal content that expressions like 'knows that there is someone else in the room' have no use to describe a circumstance, than it is when construed as having fact-claiming content. And the explanation again must be that the verbal content they appear to ordinary language philosophers to have is not their actual content. What the nature of their content is may be arrived at by looking again at the view which appears to be implied by a number of things Quine has said.

It goes without saying that no philosopher, regardless of any unconscious fantasies he may have about the magical power of words, consciously thinks that ontology, the contents of the cosmos, can be enlarged or diminished by grammatical gerrymandering with nomenclature. It would be unrealistic, to say the least, to disagree with Mill's words: 'The doctrine that we can discover facts, detect the hidden processes of nature, by an artful manipulation of language, is so contrary to common sense, that a person must have made some advances in philosophy to believe it; men fly to so paradoxical a belief to avoid, as they think, some even greater difficulty, which the vulgar do not see.'[28] And the belief that the world of things can be altered by altering grammatical categories is too absurd, too contrary to experience, to attribute to anyone who has not lost touch with reality, regardless of the extent of his 'advances in philosophy'. It is not a belief that anyone could fly to in order to avoid a difficulty, however great. When a philosopher uses language, in the expression of his position, which implies that objects can be created and destroyed by the manipulation of grammar, we are compelled to think that the 'objects' are of a very special kind: they are, we might say, philosophical entities and do not have the substantiality required to make them resistant to control by language. It would be the very height of absurdity to imagine that a philosopher, despite the suggestion of his language, might consciously believe that by reclassifying proper names like 'John' and 'Harry' with abstract nouns the nature of the bearers of the names would be changed. But ·it stretches the imagination not a whit to think that a philosopher has the idea that by grouping classes of terms in special ways he brings new entities into existence. And this is so because the 'entities' which are invoked by philosophical tinkering with grammatical categories are understood to be the kind that can be invoked in this way. The following passage helps us see what these are. In *Word and Object* Quine writes:

This chapter has been centrally occupied with the question what objects to recognize. Yet it has treated of words as much as its predecessors. Part of our concern here has been with the question what a theory's commitments to objects consists in (§49), and of course this second-order question is about words. But what is noteworthy is that we have talked more of words than of objects when most concerned to decide what there really is: what objects to admit on our own account.

This would not have happened if and in so far as we had lingered over the question whether in particular there are wombats, or whether there are unicorns. Discourse about non-linguistic objects would have been an excellent medium in which to debate those issues. But when the debate shifts to whether there are points, miles, numbers, attributes, propositions, facts, or classes, it takes on an in some sense philosophical cast, and straightway we find ourselves talking of words almost to the exclusion of the non-linguistic objects under debate.[29]

These words help us understand the nature of the mysterious entities which, unlike wombats and unicorns, are subject to control by the manipulation of grammar, and what it is about them that makes it possible to annihilate or create them by narrowing or widening grammatical categories. When the question regarding what a theory's commitment to objects *consists in* is a 'question about words' we are constrained to think that the theory is itself in some way about words. And when a philosopher perceives that his talk is more about words than about objects, even when his concern is to determine what there 'really' is, his questions as to what there really is has to be understood in terms of his perception. And understood in this way, the *philosopher's* question about what there really is, unlike the corresponding questions of the explorer and the archaeologist, is in some way about language: the reality he is inquiring into, the 'ontology' which is the subject of his discourse, is linguistic in character. As against non-linguistic objects like wombats and unicorns, the 'objects' belonging to philosophical ontology are verbal, or, if not themselves verbal, are related to terminology more like the way in which a shadow is related to the thing that casts it than like the way in which the thing is related to its name. The 'entities' which come under discussion when there is, in Quine's words, a 'shift from talk of objects to talk of words as debate progresses from existence of wombats and unicorns to existence of points, classes, miles, and the rest'[30] are linguistically engineered illusions. What comes out clearly enough to be unmistakable is that the 'shift

to talk of objects to talk of words' consists of an artificial grammatical manoeuvre in which nouns like 'point', 'class', and 'mile', which are not names of
things are assimilated into the class of nouns like 'wombat' and 'unicorn',
which are names of things. The recategorization is a piece of holiday grammar.
No corresponding semantic changes are made in the actual use of the words;
they become names of things, not in fact, but only in name. And what creates
the delusive idea that ontology is under discussion, that 'ontological slums'[31]
are being cleared out or that a domain of reality is being added to the cosmos,
is that the idle terminological classification is made in the ontological idiom,
the form of speech in which language is used to refer to extra-linguistic
objects.[32]

Wittgenstein has said that when we come upon a substantive we tend to
look for a substance. These words do not describe what actually happens and
were, perhaps, intended to call our attention, in a graphic way, to a tendency
to classify all nouns with the special group of nouns which are names of
things. This tendency, which no doubt has its roots in our earliest experience
with language, may be an echo of the time when the words we learned were
the names of things. John Stuart Mill's idea that all or nearly all words are
names represents, at one level, a hidden reclassification of words under the
term 'name', i.e., it represents an artificial use of the term stretched to cover
words which do not normally count as names. At a deeper level of the mind
it harks back to the first vocabulary in the history of our education. Be this
as it may, the insight that Wittgenstein's words give us into the nature of the
activity of philosophers who appear to look for, and to find, substances
corresponding to substantives is that it is a creative, if semantically substance-
less, manoeuvre with category words. The stretched use of the expression
'name of a thing' creates the delusive impression that a realm of objects has
come into existence; and if we consider the importance to philosophers of
the idea that ontology is in dispute, the unavoidable conclusion is that the
stretched use is introduced for the sake of the illusion it creates. The existence of the illusion is determined by the manipulation of grammatical
classifications, and this is responsible for the idea, which caters to the wish
for omnipotence of thought, that ontology is not merely discovered, but
created, by doing things with language. Philosophical ontology is a recapitu-
lation of, and indeed is created and annihilated by, academic moves with
terminological categories. It is nothing more than a language game the moves

of which are made in the ontological form of discourse. To put it aphor-
istically, the ontology of the philosopher is ontologically presented grammar.

This explanation of the position that the contents of the cosmos can be
enlarged or diminished by a 'play of words', to use Hume's expression, or by
manipulating the 'breadth of categories', to use Quine's expression,[33] afford
us a possible explanation of the nature of philosophical claims which appear
to be antithetical to ordinary language and which can be made by someone
who does not give up everyday usage. This explanation Moore would not
accept, but he would have wished to have it examined. The explanation is
that the claims are not misdescriptions of ordinary language, but present in
the ontological idiom make-believe changes in the use of everyday expressions.
For example, the philosophical view that space is unreal presents different
faces to different philosophers: to some it presents an ontological face and
seems to make an incredible factual claim; to others it presents a verbal face
and seems to make an equally incredible claim about that part of everyday
language which employs space-indicating terminology. The invisible reality
behind the visible faces is academically altered language presented in the
ontological mode of discourse. It is this behind-the-scenes game with language
which makes it possible for philosophers to say without embarrassment the
strange things that they say. The game is the reality; the faces are appearance.
Behind the game there undoubtedly lies a psychological reality, which gives
the greatest importance to the game. For philosophy, like art, is an answer to
the deep need of people 'to sugar the bitter pill of life with illusion'.[34]

NOTES

[1] The currently popular word is 'unpacking', which carries with it associations that are
unmistakable. The image the philosophical analyst creates of himself is that of someone
who digs below the surface to discover unexpected contents: he is a kind of semantic
archaeologist.
[2] 'Philosophy and "Common-Sense" ', in Alice Ambrose and Morris Lazerowitz (eds.),
G.E. Moore. Essays in Retrospect, p. 203.
[3] Introduction to Ernest Gellner's *Words and Things*, p. 13.
[4] *Philosophical Essays*, p. 131.
[5] Ibid., p. 141.
[6] *Knowledge and Certainty*, p. 177.
[7] 'A Defence of Common Sense', *Philosophical Papers*, p. 41.

[8] *Philosophical Studies*, p. 228.

[9] 'A Defence of Common Sense', op. cit., pp. 36–7.

[10] *Philosophical Studies*, p. 226.

[11] My italics.

[12] *Philosophical Investigations*, p. 47.

[13] Ibid., p. 49.

[14] What lies behind the velocity of light argument is the wish to mark the difference between seeing a physical object and 'seeing' a mental image.

[15] *The Blue Book*, p. 59.

[16] *Philosophical Investigations*, p. 43.

[17] To my knowledge, Wittgenstein nowhere says this outright; but it is certainly one of his ideas. In *The Blue and Brown Books* he states, for example, 'Ordinary language is all right' (p. 28); and in the Yellow Book he is reported as saying, 'What the bedmaker says is all right, but what the philosophers say is all wrong'. The idea which comes through here, as in other places, is that what the philosopher says about ordinary language is wrong, that he misdescribes it.

[18] This does not prevent him from also holding that entities are being postulated, rather than created, as is suggested by phrases like 'a theory's commitment to objects'. See 'On What There Is', pp. 10–11, and 'Logic and the Reification of Universals', p. 103, in *From A Logical Point of View*.

[19] 'Semantics and Abstract Objects', *American Academy of Arts and Sciences* 80, p. 93.

[20] Ibid., p. 94.

[21] Ibid., p. 93.

[22] Ibid., p. 92.

[23] This notion is given oblique expression in James S. Miller's witticism which Quine quotes on the frontispiece of his *Word and Object*: 'Ontology recapitulates philology'.

[24] Moore's expression.

[25] There is much talk these days about ordinary language being 'theory laden'. Without going into this philosophical notion, it would seem plain enough that a language in which rival theories can be expressed is not laden with or based on any of them.

[26] *Philosophical Papers* (eds. J.O. Urmson and G.J. Warnock), p. 130.

[27] *Philosophical Investigations*, p. 48.

[28] J.S. Mill, *A System of Logic*, Book II, Chapter VI.

[29] Pp. 270–71.

[30] Ibid., p. 271.

[31] Ibid., p. 275.

[32] It should be noted that Quine explicitly rejects the idea that 'the acceptance of abstract objects is a linguistic convention distinct somehow from serious views about reality'. (Ibid., p. 275) His notion is that 'The question what there is is a shared concern of philosophy and most other non-fiction genres'. (Ibid.) But what a philosopher *thinks* he is doing need not coincide with what he in fact is doing. Indeed, even the linguistic interpretation of Moore's defence of Common Sense implies that philosophers have always labored under a false idea about the nature of their work. An observation of Wittgenstein's is worth repeating here. The confusion which pervades philosophy, he is reported to have said, consists of supposing that a philosophical problem 'concerns a fact of the world instead of a matter of expression'. (The Yellow Book).

[33] *Word and Object*, p. 275.

[34] Richard Sterba, 'Remarks on Mystic States', *American Imago* 25, p. 85.

CHAPTER VII

THE SEMANTICS OF A SPINOZISTIC PROPOSITION

In Part II of his *Ethics* Spinoza professes to demonstrate the metaphysical proposition (VII) that the order and connection of ideas is the same as the order and connection of things. Taken at face value, as making a factual declaration about a correspondence which exists between our ideas of things in the external world and the things themselves, e.g., the planet Jupiter and its relations to other things or my wrist watch and its relation to its physical causes, it makes an astonishing claim, one which it is hard to think could be believed or be put forward in earnest by anyone. It is a belief we can imagine entering into a dream, but hardly one which might be held in the hard light of day. It would be strange enough if Spinoza confined the proposition to himself and made the claim that the order and connection of his ideas was the same as the order and connection of things, and thus by implication that introspective attention to his thoughts would give him knowledge of things. It is even stranger when we realize that Spinoza intended his proposition to be understood quite generally as applying to everyone, i.e., as making the claim that the order and connection of everyone's ideas is the same as that of things and therefore that each of us could obtain information about the world by merely introspectively attending to his ideas.

Spinoza took his proposition to be an immediate consequence of the axiom, laid down in Part I, that the idea of a thing which has been brought about by causes implies knowledge of its causes, but instead of being made uneasy about his inference he characterized it as obvious. Neither did his proposition appear to have raised doubts in his mind about the initial proposition from which he inferred it. Indeed, if we pause to think on it, the axiom is no less incredible than the derived proposition: translated into the concrete, it implies, for example, that anyone who has the idea of cancer knows, in virtue of having the idea, the causes that produce it. It too strikes one as a proposition which could only be embraced in a dream. Both propositions, the axiom and its putative consequence, are so flagrantly false that it is more reasonable to think that Spinoza's words do not mean what they seem to mean than that the propositions they appear to express were actually believed by him.

Regarding arguments advanced in support of philosophical propositions which go against common sense, Moore wrote: 'I think we may safely challenge any philosopher to bring forward any argument . . . which does not, at some point, rest upon some premiss which is, beyond comparison, less certain than is the proposition which it is designed to attack.'[1] Moore would certainly seem to be right. To illustrate, take Gorgias' statement that communication by means of language is impossible. If we take it to be the factual declaration it looks to be, it must be evident to everyone that there is no real possibility of a philosopher acting on the argument and rejecting a fact he lives with daily. His daily behavior shows this, as well as the fact that he presents his argument and communicates its conclusion to others. It may be puzzling, although it is easy to see, that the behavior of a philosopher who accepts the argument is in no way different from that of a philosopher who does not. Both act alike and furthermore act like people who are ignorant of philosophy. The Gorgias philosopher does not behave like someone in the tower of Babel. One of Wittgenstein's well-known remarks is that philosophy is like an engine idling, not when it is doing work; and what we observe is that a *philosophical* belief about an everyday phenomenon is idle in the everyday behavior of philosophers. Looked at from the standpoint of behavior it would seem that everyone takes for granted that any argument against the proposition that communication occurs is 'beyond comparison' less certain than is the proposition itself. What is baffling, however, is that the philosopher does not give up his argument; instead, he retains it along with his everyday belief. Both the argument and the belief it goes against maintain an amicable coexistence in his mind. Reflection on the oddness of a philosopher who, seriously and not in a moment of whimsy, argues against a proposition he has not the slightest thought of giving up, suggests the possibility that the motive of the argument, whatever it may be, is not to upset known fact; and it suggests furthermore that the words 'communication is impossible' are not intended to express the proposition they appear to express. It should be noticed that these possibilities tend to dispel the air of unreality which surrounds not only Gorgias' proposition but all paradoxical philosophical propositions, including Spinoza's; and they need more looking into than they have had so far.

Although the sentence 'The order and connection of ideas is the same as the order and connection of things' appears to make a factual declaration about things and ideas, it is not difficult to see that its use in Spinoza's work is not to express an empirical proposition, one to the determination of whose truth-value experience is relevant. For the sentence is not linked with any kind of empirical investigation. There is no hint whatever of having recourse to a series of observations or of checking in some way our ideas of things against the things, in the usual ways that this is done in everyday life, as when, for example, we wish to determine whether we are actually seeing what we seem to be seeing. It would be unrealistic to imagine that an enduring oversight had occurred. We have, instead, to think that what the sentence states is not connected with observational procedures, and that the reason for this is that its use is not to express a proposition to which observation is relevant.

A philosopher who might possibly agree with this assessment of the Spinozistic utterance and give up the notion that it presents an empirical claim, may nevertheless not give up the notion that it does actually make a claim about the relationship between ideas and things. He might maintain that it expresses a proposition about things and ideas, but one which possesses its truth-value by logical necessity. Many philosophers seek refuge from the insecurity of their subject in the Kantian doctrine that in addition to synthetic, empirical propositions, which are about things, and analytic propositions, which are not, there are propositions that both have factual content and are characterized by the kind of necessity that characterizes analytic propositions. As is known, Kant attempted, with the help of his category of synthetic *a priori* propositions, to rid philosophy of its intellectual anarchy. His own creation, however, is beset by the same anarchy of opinions, which hardly makes it the Atlas on whose shoulders Philosophy could find a secure resting place.

It is important to re-examine with care the notion of a synthetic *a priori* proposition, but what needs to be looked into here is the claim that such a proposition presents factual information. It is usually held that an analytic proposition, one whose negation is or implies a self-contradiction, embodies no factual information, and it is easily seen that neither Spinoza's statement nor the axiom from which he infers it is analytic. The sentence 'The order and connection of ideas is not the same as the order and connection of things' does not express a proposition which implies a contradiction of the form $\phi \cdot \sim\phi$, and neither do the words 'has an idea of the moon but has no knowledge of its causes'. Instead, so a Kantian defender of philosophy mould urge, they express non-self-contradictory *logical* impossibilities, impossibilities of thought which are also impossibilities of things. The affirmative propositions corresponding to these impossibilities are of the form 'It is necessarily the case that . . .', and they are necessities of thought and of things. Unlike the statement that the outermost planet of the solar system is smaller than Saturn, which is about things and is such that its denial is conceivable, the two statements of Spinoza, on a Kantian explanation, are about things and are such that their denials present impossibilities of thought.

It can readily be seen why a proposition that is true by *a priori* necessity is not open to any sort of empirical verification procedure. The reason is that observation and experiment are possible forms of investigation in those cases only where the opposite of what they establish remains conceivable and might, instead, have been the outcome of the investigation. This is out of the question in the case of a proposition the negation of which yields a logical impossibility. We cannot by observation establish the falsity of a proposition which implies a self-contradiction or is in some way logically impossible, since we do not know what it would be like to observe what we cannot conceive. I can imagine myself observing a state of affairs which does not in fact exist, for example, there being a rhinoceros in my study, because I can imagine there being a rhinoceros in my study. But there is no imagining or conceiving myself observing a rhinoceros that is larger than it is, because there is no conceiving or imagining a rhinoceros that is larger than it is. A proposition that is false by logical necessity cannot, in principle, be shown false by experience. Hence, the corresponding *a priori* true proposition, 'There is no rhinoceros that is larger than itself', or the equivalent proposition, 'All rhinoceri are such that they are not larger than themselves', cannot in principle be verified by experience. Wittgenstein has remarked that 'Tautologies and contradictions are not pictures of reality. They do not represent any possible situations.'[2] He might have included synthetic *a priori* propositions along with tautologies, and logical impossibilities with contradictions, as not being 'pictures of reality'. Put generally, a proposition which possesses its truth-value *a priori* fails to be a 'picture' of an actual or theoretical situation, and this is why observation, the examination of things, plays no role in determining its truth-value.

We can see that *a priori* propositions are not 'pictures' of what there is or of what there might be by noticing that the consideration for saying that an *a priori* true proposition is not open to testing by observation is also a consideration for saying that such a proposition makes no declaration about things. The point on which the argument hinges is that if the denial of a proposition, ~*p*, is not about things, then *p* is not about things. The proposition that the outermost planet of the solar system is smaller than Saturn makes a declaration about planets, as does its denial, that the outermost planet of the solar system is not smaller than Saturn. Each denotes a conceivable state of affairs. A necessarily true proposition, for example, that a pain is a feeling, is in a logically different case. Its denial presents us with an inconceivability, a conceptual blank, not something that is conceivable although incredible: the phrase 'pain that is not a feeling' has no use to characterize a pain. A. Heyting, who was concerned with the validity of proof of *reductio ad absurdum*, said that the supposition that a square which at the same time is a circle has been constructed 'has no clear sense, because it can never be realized'.[3] Regardless of whether the hypothesis has a clear sense, what we can say is that the hypothesis that a square circle has been constructed has no *descriptive sense*. It is not hard to see why the hypothesis 'can never be realized'. The reason is that it is not about a geometrical figure. The expression 'round square' does not have a use to refer to a figure and the words 'A square which at the same time is a circle cannot be constructed' is not about a figure which geometers for one reason or another are unable to construct. If the sentence described a figure, the figure *could* in principle be constructed.

To put the matter shortly, the negation of a proposition which is necessarily true is a logically impossible proposition, one which because of its nature there is no conceiving. It therefore is not about what a thing is not. Instead, it is not about a thing. Neither can the corresponding *a priori* true proposition be about what a thing is. For since its denial does not say what a thing is not, it cannot say what a thing is. An *a priori* true proposition, regardless of whether or not it is synthetic, makes no declaration about things. Thus, construing Spinoza's statement that the order and connection of ideas is the same as the order and connection of things to be synthetic *a priori*, rather than analytic, does not save it from the charge that it is not about how ideas and things are related to each other: it is not about things and ideas. The appearance the proposition has of being about the world is delusive; and we cannot imagine that it would hold the interest of philosophers, both those who reject it as well as those who accept it, if they did not believe the proposition to be what it appears to be. The powerful hold the appearance has on the minds of philosophers suggests the possibility that the function of the appearance is comparable to that of a dream. A dream enables the dreamer to go on sleeping; and we may guess that the belief that the proposition is about things enables the philosopher to remain oblivious to processes going on in his own mind.

We get a better view of the nature of Spinoza's proposition, if we consider the English *sentence* which expresses it. What G.E. Moore has said regarding the word 'good' is pertinent in this connection: '. . .my business is not with its proper use, as established by custom. . . My business is solely with the object or idea which I hold, rightly or wrongly, that the word is generally used to stand for. What I want to discover is the nature of the idea or object.'[4] Philosophers frequently protest that their investigations are not concerned with usage, or, at any rate, that verbal usage is not an important part of their investigations. One thought which this denial provokes is that philosophers must suffer from the uneasy feeling that their work *may in some way* be verbal. Not to go into this for the moment, our first business is to determine how terminology is being used in the Spinozistic sentence, whether it is being used as established by custom or whether it violates customary usage, or whether, to use Hume's expression, the sentence presents a play of words. For an examination of the proposition the sentence is normally taken to stand for has led us into a *cul de sac*: we do not know what to make of a proposition that appears to be factual but is not empirical. Only after we have arrived at a correct understanding of the sentence will it become our business to analyze 'the object or idea', that is to say, the proposition, which the sentence is used to express. It should be pointed out that the interpretation which is finally placed on the sentence will deserve consideration only if it throws light on two things, amongst others. One is the appearance the sentence presents of expressing a proposition about the connection between things and ideas, or, perhaps better, the air the sentence has of imparting factual information. The other is the question how anyone could accept, as Spinoza and others seemed to have done, the incredible claim apparently made by the sentence. Bishop Berkeley's well-known rule was to speak with the vulgar, and think with the learned. It would seem that Spinoza's rule was to think, sincerely, the utterly incredible but speak and act with the sober.

The answer to the question as to what the sentence is about, what it says, is perhaps best arrived at by considering the following remarks taken from an American Bar Association account of a series of law cases involving collision:

Gordon parked his car next to an apartment house, unaware that a burglary was taking place inside the building. The burglar leaped from a second story window, landed on top of Gordon's car, and made good his escape. The car was damaged by the burglar and needed $180 worth of repairs. Was he entitled to collect insurance?

The insurance company contended that he was not covered. It told him: 'True, you have collision insurance, but this was not a collision.' However, when Gordon took the matter to court, the judge ruled in his favor. 'Collision', the judge argued, 'is the impact of objects brought about by one of them striking the other, and the occurrence fitted this description.'

Typicallly, collision insurance covers damage done to one car when it bumps into another. However, it has often been held to extend to less typical cases of impact. One motorist collected collision insurance when his car struck a mail box by the side of the road, another when his car rolled into an open elevator shaft and fell to the bottom, another when his car rammed into the curb.

Nevertheless, there are limits. In one case, the paint on a man's car was damaged when he ran into a hailstorm. He insisted that this was covered by his collision insurance, because his car had 'collided' with the pellets of ice. But the court decided that this was stretching language too far. Dismissing the claim, the court said: 'We do not speak of falling bodies, such as sleet or hail, as colliding with the earth. In common speech, an apple is said to fall to the ground, not to collide with it.'

This report makes two things stand out with great clarity: how talk about words can look like talk about things, and how disagreement about language alterations can look like disputes over correct usage, which in turn can look like disputes about what there is or is not. The insurance company's contention that the damage to Gordon's car was not the result of a collision looks like a statement about the facts of the case, namely, that there had been no collision. The judge's decision that a collision had occurred also has the appearance of being a statement about the facts of the case. It is plain, however, that the opposing opinions were not based either on misinformation or on inadequate information regarding what had happened. No further piece of factual information could have the effect of making either the judge or the insurance representative change his mind and declare, 'If I had known this, I would not have said that there had been (or that there had not been) a collision.' All the material facts of the case that are relevant to removing the difference between them are known, without producing agreement. The reasonable inference to draw is that the conflicting opinions are not about what had occurred. Regardless of the forceful impression to the contrary, which is not easy to dismiss, the words 'The thief collided with Gordon's car' are to be understood as saying nothing about a collision between the thief and the car; as are also the words 'The thief did not collide with it'.

Once it is realized that the opposing opinions are not about what had happened, there is an inclination to say that they are about the correct use of 'collision', i.e., that they are verbal statements which are implicitly, if not explicitly, about the customary, conventional use of the word. On this reading of the nature of the disagreement, the assertion that a collision had taken place becomes the assertion that the term 'collided with' applies correctly to what is described by the expression 'landed on', and the assertion that none had taken place becomes the assertion that the term does not correctly apply to the described case. But this reading is as unsatisfactory as the first one. For the actual use in the language of the word 'collision' is known equally with the material facts of the case. The company which provided the insurance and also the judge who rules against the company knew perfectly well whether, as a matter of standard usage, 'collided with' applies to whatever 'landed on' correctly applies to, or, to put it differently, whether the expression 'landed on but did not collide with' has an actual use to describe an occurrence.

The disagreement is in some way verbal, but it is not a disagreement about what is dictated by correct usage. In what way the disagreement is verbal becomes clear if we consider the judicial decision against the driver whose car was damaged by hail, and compare it with the judgment that the car which had fallen down an elevator shaft had suffered damage in consequence of colliding with the bottom. The judge whose decision went against the man's claim of damage to his car by colliding with falling hailstones gave as his reason that it was *stretching* language too far to apply 'collide with' to a case of hailstones striking a car. The implication of this judgment is that the phrase 'collided with the hailstones that fell on it' is not being used either to describe what had happened or to exhibit the correct use of 'collided with'. Rather, the implication is that the plaintiff's claim presents a semantically doctored application of 'collided with'. Instead of being introduced in the form of speech in which an expression is mentioned, the piece of stretched language is introduced in the indicative, nonverbal form of speech which we use to talk about things and occurrences. This tends to conceal what is being done with nomenclature: linguistic innovation is obscured by the facade of the ontological form of language.

It is instructive to compare this decision with the decision to count the car's falling down the elevator shaft as an instance of one object colliding with another. The judge who handed down this opinion apparently did not think that language was being stretched too far in *this* case, although we should continue in common parlance to speak of apples and hailstones falling to the ground, and not of colliding with it. It is easy to see how the two judicial opinions could be turned into an irresolvable antinomy, one in which neither side need ever give way to the other in consequence of a new fact being brought forward. For no sort of fact is in question. There is only a conflict over tinkered usage, and what determines its acceptance or rejection need be no more than preference, although in ordinary cases practical considerations, such as being paid for physical damage, play a role. In philosophy, which as Kant said bakes no bread, only preference, which can give way to another preference, determines the acceptance or rejection of re-edited terminology.

Wittgenstein remarked that a philosopher rejects a notation under the impression that he is upsetting a proposition about things. Philosophers also adopt notations with the thought that they are making scientific statements about the world, and they dispute each other's claims under the illusion that they are in disagreement about the existence and nature of things. According to this notion, which can hardly be expected to fill philosophers with enthusiasm, the philosophical disagreement over Spinoza's proposition, or the similar disagreement over the Parmenidean view that the thinkable and the existent coincide, is like the disagreement about whether or not a falling apple collides with the ground when it strikes it. The Parmenidean view, instead of being either about how thought and reality are related to each other or about how the terms 'exists' and 'thinkable' function with repsect to each other in the language, presents a stretched use of 'exists', a use which by fiat dictates its application to whatever 'thinkable' applies to. In the special Parmenidean language, which it is to be noticed never intrudes itself into the language of everyday talk, the expression 'thinks of what does not exist' has no use. The philosophical sentence 'Thought and existence coincide precisely' is not used to state a putative fact about thought and things, and so is secure against the possibility of being upset by recourse to the facts. Neither is it used to put forward a claimed fact about actual usage, and so is secure against the possibility of being upset by recourse to usage. It is irrefutable because it does no more than present redistricted terminology; and it does this in the form of language in which we speak of things and occurrences, as in the collision cases. The difference between the Parmenidean proposition and the collision claim and counterclaim is that practical matters hang on the outcome of the latter, i.e., on which is adopted, and none on the adoption (or rejection) of the former. The Parmenidean language-innovation is not linked to a practical job. Its work is to bring an illusion into existence. And probably there is more to it than this. Early in his work Freud found that dreams and fantasies, as well as neurotic symptoms, are multidetermined and dynamically linked with a number of underlying ideas, all of which are beyond the reach of the conscious part of the mind. It is hard to think that terminology that is revised for its fantasy-value is not put into the service of deeper mental needs. As with a dream or a painting, what we are aware of in the contemplation of a philosophical theory is not all that is present in the mind.

How we are to understand the Parmenidean view, in part, is suggested by Aristotle's observation that man by nature desires knowledge and that it is because of wonder that men begin to philosophize. Science undoubtedly began in and has been sustained by curiosity, and is pursued not only for its material benefits but also for the sake of understanding the world. The philosopher, unlike the scientist who goes to things in order to satisfy his curiosity about them, remains aloof and prefers to be an Olympian whose gaze is fixed on concepts. An explorer of the world who conducts his explorations in thought alone is confined to gratifying his wish to scientific knowledge in a substitutive way, by settling for a contrived appearance of science in place of science itself. The Parmenidean utterance, which does nothing more than covertly present rearranged application of 'thinkable' and 'exists', generates the idea that the utterance expresses a basic theory about the connection between thought and what there is. It is this idea which has to satisfy the philosopher's thirst for cosmic knowledge.

The claim has been made for philosophy that it is the mother of the sciences, and that although herself sterile of scientific results nevertheless has given birth to disciplines which are productive of results. The sober truth is that a science is born when interest in playing magical tricks with words is supplanted by interest in the phenomena the words are normally used to refer to. Thus, a philosopher like Zeno invents a game which can be played with 'motion', 'rest', and related words and produces nothing more solid than an illusion and endless disputation; whereas a nonphilosophical scientist like Galileo studies the phenomena referred to by these words and erects a structure of theoretically important as well as practically useful facts about the world. In his famous parable of the cave, Plato described the nonphilosopher as someone who sees only shadows which he takes for reality. Very likely the correct interpretation of this is that in a distorted way it pictures the philosopher rather than the nonphilosopher, and stems from a glimpse Plato had of the true state of affairs within himself. Behind the lofty picture of himself as a voyager who returns to the world of shadows from the supersensible realm of abstract universals — where 'abides the very being with which true knowledge is concerned; the colourless, formless, intangible essence, visible only to mind'[5] — he had a perception of himself in flight from reality. The forlorn image of the philosopher as someone who turns his eyes away from the real world and prefers to look at semantic shadows could be successfully denied with the help of the defense mechanism of projection, i.e., by displacing this image onto the nonphilosopher: it is the nonphilosopher, not I, who lives in a world of shadows.

Spinoza's words, like those of Parmenides, are to be construed as introducing amended nomenclature for the consolation of an illusion. What Spinoza put forward when he wrote that 'The order and connection of ideas is the same as the order and connection of things', is a special, doctored use of 'same as', which dictates its application to what the two expressions 'the order and connection of ideas' and 'the order and connection of things' are used to apply to. The actual, everyday use of 'same as' is not changed; the new application has only a holiday purpose, which is to say that its work is to fabricate a fantasy. The axiom from which Spinoza professed to derive the theorem has also to be viewed as introducing, under cover of the ontological form of speech, academically re-edited terminology, the axiom, namely, that knowledge of an effect depends on and involves knowledge of its cause. Superficially, it appears to make an entailment claim, to the effect that knowledge of an occurrence E which is brought about by a cause *entails* having

knowledge of its cause. Taken as such, the claim is that it is logically impossible both to have knowledge of E and to lack knowledge of what its cause is.
And what this comes down to, when restated in terms of actual usage, is that
the expression 'has knowledge of an occurrence but no knowledge of its cause'
has no descriptive function in the language.

Now, it does in fact make literal sense to say, e.g., 'Spoonerisms are a well-
known psychological phenomenon but what their cause or causes are has yet to
be discovered', and 'Cancer is a much studied disease but some of its causes are
unknown'. Neither is it true that actual usage dictates the application of
'knows the cause of occurrence E' to whatever 'has knowledge of E' correctly
applies to. The Spinozistic philosopher, as expected, lives with the same
linguistic realities we live with, and knows them as well as anyone else. Thus,
when in his philosophizing he seems to flaunt them, instead of attributing on-
and-off amnesia to him we have to credit him with a kind of verbal wayward-
ness which in no way conflicts or interferes with his normal use of language.
The sentence 'Knowledge of an occurrence entails knowledge of its cause'
does not serve to communicate either a fact about the world or a fact about
language. Instead, it brings before us a linguistic work of art, one which caters
to a fantasy-need rather than to a work-a-day demand of life. By an act of
linguistic fiat the phrase 'knows the cause of E' is made to apply to whatever
'has knowledge of E' is used to apply to; and because this innovation is
presented in the form of speech used to refer to facts of the world, the
Spinozistic sentence enables us to enjoy a false belief.

If the psychoanalytic account of the structure and dynamics of the mind
is correct, it is to be expected that there is more to a philosophical theory
than has so far been brought to light, that indeed the greater part of its work
is carried on in the less well lit part of the mind. Gorgias is said to have remarked that with regard to theatre the deceived are wiser than the not-
deceived, and it would seem that philosophers wish to have this kind of
wisdom about their discipline. For they are dupe to a deception which they
protect with every means in their power. Put in terms of the analogy of the
mind to an iceberg, the semantic theatre created by Spinoza's utterance is
only the tip of the iceberg, its visible part. Below this, under the surface and
hidden from conscious view, is a piece of contrived semantics which engin-
eers the illusion. Still lower, buried in the darkest recesses of the mind, lies
a constellation of rejected thoughts which the philosophical sentence is made

to express. This third component in the structure of a philosophical theory not only is the source of the enchantment the theory has for us, it is also the main cause of the blindness to the game played with words. Wittgenstein has observed that the words of a philosophical sentence could be used to express a factual proposition.[6] It is this which makes it possible for a philosophical pronouncement to create a lively, if erroneous, impression, and also for it to give secret expression to a wished-for state of affairs.

What the sentence 'The order and connection of ideas is the same as the order and connection of things' is made to express for the subterranean part of the mind is a matter of educated guesswork, and a guess is permissible even in the absence of professional credentials. The empirical picture that is naturally associated with the sentence provides a possible clue to its underlying purport. This picture, which resembles the one we tend to associate with Leibniz's famous statement, 'The microcosm mirrors the macrocosm', is that what is in the mind duplicates what is in the world and that we can know what there is in the world in detachment from the world. Psychologically, this detachment represents a turning away from the world:[7] it is as if Spinoza had said: 'I have no need of you; I am self-sufficient, and can fall back on my own resources'. This attitude he took not only to the Amsterdam community which had excommunicated him, but to the entire world, and so made himself secure against any further excommunications. In his personal life he became a near recluse, and in his unconscious he seems to have become a cosmic hermit. Adopting the attitude of self-sufficiency is bound up with a heightening of one's narcissism, and in Spinoza's case it took the form of fantasied omniscience, the belief that he had the power within himself to know all that there is. The metaphysical system he developed in his *Ethics*, which presents a comprehensive map of the Universe, is a working out of his imagined omniscience; and it compensated for the dreadful curse with which he was excommunicated. Interestingly enough, Spinoza's theology, in which God is the eternal and all-encompassing reality, made it psychologically impossible for him to suffer expulsion: for everything that exists is in God, including ourselves. Hence in God's love of Himself, some measure of His love is ours. We may guess that Spinoza's theology represents in part a wish to return to the undisturbed security of the place which is the first home of everyone, the womb.

NOTES

[1] 'Some Judgments of Perception', in *Philosophical Studies*, p. 228.
[2] *Tractatus Logico-Philosophicus*, 4.462. Pears and McGuinness translation.
[3] His account of Griss' position in *Intuitionism. An Introduction*, p. 120
[4] *Principia Ethica*, p. 6.
[5] *Phaedrus*, sec. 248 of B. Jowett's translation of *The Dialogues of Plato*.
[6] See *The Blue Book*, pp. 56–7.
[7] Spinoza's friend, Jarig Jelles, said that 'he withdrew himself entirely from the world and hid himself'. From A. Wolf (ed.), *The Oldest Biography of Spinoza*. Quoted by Lewis Samuel Feuer in *Spinoza and the Rise of Liberalism*, p. 140.

CHAPTER VIII

THE INFINITE IN MATHEMATICS

In their everyday, popular use the words 'finite' and 'infinite' are connected with the idea of quantity: the first with the idea of a limited quantity or amount, the second with the idea of the unlimited and the vast. Thus, in everyday talk something is said to be infinitely far from us, e.g., a remote galaxy, when it is a vast distance away, in contrast to something which is said to be far but not infinitely far; and infinite wealth, as against limited assets, is understood to be enormous wealth. If we were told that Jones had infinite credit at the bank, we should naturally infer that he had more credit than people who had only limited credit. The words 'finite' and 'infinite' are not used in their popular senses in mathematics, but the ideas of the huge and the less than huge seem, nevertheless, to be in the background of the thinking of at least some mathematicians. One mathematician inadvertently revealed this in the lapse shown in the following words: 'Representation of a complex variable on a plane is obviously more effective at a finite distance from the origin than it is at a very great distance.'[1] This brings to mind a description in a brochure about a marshland in Ohio, as 'almost endless'.

Wittgenstein is said to have remarked in lectures that 'The idea of the infinite as something huge does fascinate some people, and their interest is due solely to that association, though they probably would not admit it'. There is reason for thinking that this remark applies to many mathematicians who adopt Cantor's notion of the 'consummated infinite', the notion that, for example, the series of numbers $1, 2, 3, 4, 5 \ldots$ form a completed totality of elements, just as the first forty odd numbers form an entire set of numbers or just as the chimney pots in Bloomsbury make up a whole class of objects in London. He is also reported as having said: 'When someone uses the expression '$\aleph_0$ plus 1' we get the picture of 1 being added to something. If I speak of 'the cardinal number of all the cardinal numbers' all sorts of expressions come to mind – such as the expression 'the number of chairs in this room'. The phrase conjures up a picture of an enormous, colossal number. And this picture has charm.' Wittgenstein declared, as did Gauss, that infinity has nothing to do with size, and there is reason to think that transfinite arithmetic is primarily a semantic creation for representing the mathematical infinite as the colossal. It may turn out that despite its containing solid mathematics, the discipline which Hilbert described as the paradise created by Cantor and which Poincaré characterized as a disease from which mathematics will eventually recover is at bottom a semantic contrivance for the production of an illusion.

Many people take the view that all number are *finite* numbers, and Russell gives one explanation as to why they do this. The phrase 'finite number' means, according to him, '0 and 1 and 2 and 3 and so on, forever – in other words, any number that can be obtained by successively adding one. This includes all the numbers that can be expressed by means of our ordinary numerals, and since such numbers can be made greater and greater, without ever reaching an unsurpassable maximum, it is easy to suppose that there are no other numbers. But this supposition, natural as it is, is mistaken'.[2] Russell's idea is that it is natural to think the finite numbers are the only numbers, and Zeno gives an argument for this notion. One thesis of Fragment 3 goes as follows: 'If there is a multiplicity of things, they necessarily are as many as they are, and not more or fewer. If they are exactly as many as they are, then they will be finite in number.' The counter-thesis is that they must also be infinite in number, but that is not to the point here. Zeno's argument amounts to the contention that a number of things must be a definite number (though we may not know what it is) and therefore finite, i.e., a number expressible by one of our 'ordinary' numerals.

In Russell's opinion the weak point in the thesis that 'if they are just as many as they are, they will be finite in number' is that it is based on the assumption that definite infinite numbers are impossible.[3] It is by no means clear what is meant by the phrase 'definite infinite number'. The idea of a *definite* infinite number would seem to be that of a number that is like one denoted by an 'ordinary' numeral but greater than any such number. This may well strike one as being a number that is finite, (or, to use Galileo's word, 'terminate') but too great to be finite, a *finite infinite* number. The phrase 'the consummated infinite' also suggests the notion of an infinity of elements whose number is made definite and thus finite. Be this as it may,

Russell's objection brings to our attention an important point about the words 'All numbers are finite': this is that if the word 'finite' has a correct application to numbers, then its associated antithetical word 'infinite' must also have a correct application to numbers. It *may* be that in fact the application of 'finite' to numbers does not represent a correct use of the word: it *may* be that the sentences '5 is a finite number' and 'a thousand billion is a finite number' do not represent a correct use of the word. But if they do, then 'infinite' must also have a correct application to *some* numbers. For otherwise 'finite' would not have a use to distinguish between numbers, set off some from others, as do, for example, the terms 'prime number' and 'proper fraction'. Without its antithesis, 'infinite number', the expression 'finite number' would contain a word which serves no function, and 'finite number' would have no use different from that of the word 'number'. It is a curious feature of the view that all numbers are finite that it rests on a distinction which at the same time it obliterates. Without a distinction between finite and infinite numbers, Zeno's putative demonstration could not even have been formulated. But if his first thesis is in fact demonstrated, there could *in principle* be no infinite numbers, which would imply that there is no distinction between finite and infinite numbers.

Wittgenstein has pointed out that in philosophy words are often used without an antithesis, in what he described as a 'typically metaphysical way'.[4] He conceived his task as being 'to bring words back from their metaphysical to their everyday usage'.[5] If, allowing that 'finite' has a correct application to numbers, we preserve the distinction implied by its use, we have to 'bring back' into usage the application of 'infinite' to numbers. Doing this upsets the thesis that *all* numbers are finite and would seem to open the way to the claim that there are numbers which are not expressible by any of the 'ordinary' numerals. Russell has remarked: 'When infinite numbers are first introduced to people, they are apt to refuse the name of numbers to them, because their behavior is so different from that of finite numbers that it seems a wilful misuse of terms to call them numbers at all'.[6] But if it is not a misuse of terminology to call 5 a finite number, it cannot be a misuse of terminology to call an infinite number a number, regardless of how it may differ from finite numbers. And of course it would be a wilful misuse of terminology to deny that the number of primes is infinite.

It is a curious and striking thing about the idea of the infinite that in some connections it would be considered unnatural to deny that 'infinite number' has a correct application, while in other and comparable connections it is natural to deny this. There would be no temptation to deny that there is an infinite number of numbers from 1 on. But it seems entirely natural to state (as the cosmological argument does) that it is in principle impossible for there to be an infinite series of causes. No one could take exception to the statement that the geometric series $1 + \frac{1}{2} + \frac{1}{4} + \ldots$ is an infinite series, as against one of its parts, e.g., $1 + \frac{1}{2} + \frac{1}{4}$; but many would take exception to the claim that an infinite series forms a whole. Again, no one would say that $\frac{1}{2}, \frac{3}{4}, \frac{7}{8}, \frac{15}{16}$ is the entire series generated by $\frac{2^n - 1}{2^n}$. But many people would, nevertheless, deny that there is such a thing as an entire unending series, and therefore would deny that '1 is *beyond*[7] the whole of the infinite series $\frac{1}{2}, \frac{3}{4}, \frac{7}{8}, \frac{15}{16}, \ldots$'.[8] Russell asserted that the first infinite number is 'beyond the whole unending series of finite numbers',[9] and went on to remark that it will be objected that there cannot be anything beyond the whole of a series that is endless. What is the difference between speaking of an infinite series and speaking of the whole infinite series, which makes people accept the one and dispute over the other? No one would be tempted to reject the statement that there is an infinite number of natural numbers or that there is an infinite number of primes; and it is mystifying that people who are introduced to infinite numbers are likely to refuse to apply the word 'number' to them. What, we may ask, is the difference between allowing that there is an infinite number of natural numbers and allowing that there is a number which is infinite?

Russell offers the following explanations:

... the number of inductive numbers is a new number, different from all of them, not possessing all inductive properties. It may happen that 0 has a certain property, and that if n has it so has $n + 1$, and yet that this new number does not have it. The difficulties that so long delayed the theory of infinite numbers were largely due to the fact that some, at least, of the inductive properties were wrongly judged to be such as *must* belong to all numbers; indeed it was thought that they could not be denied without contradiction. The first step in understanding infinite numbers consists in realising the mistakenness of this view.[10]

In philosophy we are not strangers to 'mistakes' which have the quality of elusiveness: they strongly impress some people as being mistakes, while making no such impression on others. The 'mistake' of thinking that certain of the inductive properties, such as that $n + 1$ is greater than n, are properties of all numbers, would not be accepted as a mistake by anyone who is competent to express an opinion. Russell has remarked that the 'astonishing difference' between any number occurring in the sequence 1, 2, 3, . . . and the number of all numbers in it is that 'this new number is unchanged by adding 1 or subtracting 1 or doubling or halving or any of a number of other operations which we think of as necessarily making a number larger or smaller. The fact of being unchanged by the addition of 1 is used by Cantor for the definition of what he calls 'transfinite' cardinal numbers'.[11] And one mathematician has spoken of the 'crude miracle' which 'stares us in the face that *a part of a set may have the same cardinal number as the entire set*'.[12]

As is known, Leibniz was not unaware of the miracle which Cantor later performed, except that he called it a contradiction. 'The number of all numbers', he declared, 'implies a contradiction, which I show thus: To any number there is a corresponding number equal to its double. Therefore, the number of all numbers is not greater than the number of even numbers, i.e., the whole is not greater than its part.'[13] What Russell calls a 'new' number with astonishing properties Leibniz calls an impossible number because of these astonishing properties. And the fact of being unchanged by the addition of 1, or the fact of the whole not being greater than its part, which Cantor used to define the term 'transfinite number', was used by Leibniz to deny there could be such a number. If he had been shown Cantor's symbol for the first transfinite number, '$\aleph_0$', it is fair to suppose he would have said that it is not the name of a possible number, and thus that it is not actually the name of a number. It will be remembered that Gauss protested against the use of the actual or completed infinite as something which is 'never permissible in mathematics'. The important thing to notice about this disagreement is that it is not the result of incomplete knowledge of the facts on the part of anyone. If it is a fact that $n + 1$ is greater than n for *any* number, then all the parties to the disagreement know this; and if it is not a fact, then this too is known. There is no disagreement over whether every number n has a double, $2n$, and over whether every number n has a square, n^2, but there is disagreement over whether facts like these show that there is a contradiction in the idea of an infinite totality of numbers or whether it brings to light a characteristic of such a totality. And nothing new can be brought in to help us decide one way or the other.

Galileo gives a different answer to the question about infinite numbers. Leibniz and others think that infinite numbers are impossible, the implication being that there are none. Cantor and others think that there are such numbers and that, unlike finite numbers, an infinite number can be equal to a proper part of itself, and also that some infinite numbers are greater than other infinite numbers. Galileo's position is that such terms as 'greater than', 'equal to', and 'less than' are not applicable to infinite numbers. From the fact that there cannot be fewer squares than there are numbers of which they are the squares, i.e., than 'all the Numbers taken together', he does not conclude that there can be no infinite numbers or that infinite numbers have paradoxical properties which defeat our understanding. In the first of his *Dialogues on Motion*, Salviati asserts: 'These are some of those Difficulties which arise from Discourses which our finite understanding makes about Infinities, by ascribing to them Attributes which we give to Things finite and determinate, which I think most improper, because those Attributes of Majority, Minority, and Equality, agree not with Infinities, of which we cannot say that one is greater than, less than, or equal to another.' Against Galileo's solution Russell has the following to say: 'It is actually the case that the number of square (finite) numbers is the same as the number of (finite) numbers'.[14] What makes this 'actually the case', we are entitled to ask, as against Galileo's conclusion that 'equal to' is not applicable to the number of numbers and the number of squares?

According to some philosophical mathematicians, commonsense thinkers (as well as mathematicians who might side either with Leibniz or with Galileo) have been taken in by the maxim that if the elements of one set, α, are some only of all the elements of another set, β, then α has fewer elements than β has and β more elements than α. Russell writes:

This maxim is true of finite numbers. For example, Englishmen are only some among Europeans, and there are fewer Englishmen than Europeans. But when we come to infinite numbers this is no longer true. This breakdown of the maxim gives us the precise definition of infinity. A collection of terms is infinite when it contains as parts other collections which have just as many terms as it has. If you can take away some of the terms of a collection, without diminishing the number of terms, then there are an infinite number of terms in the collection. For example, there are just as many even numbers as there are numbers altogether, since every number can be doubled. This may be seen by putting odd and even numbers in one row, and even numbers alone in a row below:

$$1, 2, 3, 4, 5, ad\ infinitum$$
$$2, 4, 6, 8, 10, ad\ infinitum.$$

There are obviously just as many numbers in the row below as in the row above, because there is one below for each one above. This property which was formerly thought to be self-contradictory, is now transformed into a harmless definition of infinity, and shows, in the above case, that the number of finite numbers is infinite.[15]

The idea this passage tends to produce in one's mind is that a popular belief is being exposed as nothing more than a superstition, that a plain fact is being stated, and that something is being shown. Thus, the phrases 'breakdown of a maxim', 'there are obviously just as many even numbers as both odd and even numbers', 'shows that the number of finite numbers is infinite' create the impression that a proposition is being upset and that a truth about infinite sets is being held up. Parenthetically, it is worth noticing that Galileo's claim that 'equal to' does not apply to infinities is pushed aside with the words 'there are obviously just as many. . .'. Galileo certainly was not unaware of the obvious fact that every number has a square, in the face of which he held his own view. If we can ward off the hypnotic effect of Russell's words, supported as they appear to be by the actual transfinite arithmetic developed by Cantor and others, what we see is not that a position is being shown true and rival positions false. What we see, at first glance at least, is that one position is being arbitrarily embraced and that rival positions, rather than being shown false, are simply dismissed. What is called the 'breakdown' of a commonly accepted maxim turns out to be merely a rejection. And we may wonder what the nature of the 'transformation' is which consists of changing a property that is thought by some to be self-contradictory into 'a harmless definition of infinity'.

Russell speaks of 'those who cling obstinately to the prejudices instilled by the arithmetic learnt in childhood'.[16] And R.L. Wilder tries to reassure those who may feel uneasy about the notion of actual infinite numbers that the symbol '$\aleph_0$' for the first transfinite number will with practice come to have 'the same significance for us as the number 15, for example'.[17] If we take into account the difference between numbers and the numerals which denote them, or between symbols and their meanings (to which Wilder himself calls attention[18]), then what we are being assured about is that with practice we will think of '$\aleph_0$' as the name of a number, or, better, that along with '15' we will come to think of '$\aleph_0$' as a numeral. The idea which cannot fail to cross one's mind is that Cantor christened the infinite and that followers of his are trying to assure us that there really is an infant. The question is whether $\aleph_0$ *is* a cardinal number, one which gives the *size* of a collection, and whether the *actual use* of '$\aleph_0$' is to denote a number which in any way is comparable to the number 15.

As is known, not all mathematicians accept the notion of a consummated infinite. One mathematician has described an opposing position in the following way: 'Some intuitionists would say that arbitrarily large numbers can perhaps be constructed by pure intuition, but not the set of all natural numbers.'[19] A metaphysical haze surrounds talk of 'constructing numbers in pure intuition', but undoubtedly what it comes down to is talk about the ability to think of various numbers. Some people are able to think of greater numbers of objects than are others. Some arithmetical prodigies are able to do enormous multiplications in their heads, and probably can consciously and all at once entertain a large array of natural numbers. But no one, according to the intuitionist claim, has the capacity for presenting to himself the set of all natural numbers. Wilder has written:

If we analyze the psychology of the 'intuitive meaning' of the number 2, we shall probably conclude that '2 apples' brings up to the mind of the hearer an image of a *pair*, here a pair of apples. A similar remark might hold for the phrase '20 apples'; but it would hardly hold for '200 apples'. From the psychological viewpoint, it seems probable that 200 is simply one of the numbers one ultimately gets at by starting with the numbers whose mental images are distinct – 1, 2, 3, – and applying consecutively the operation of adding 1, as taught in the elementary schools. (This is certainly the case with a number like 3,762,147; it is conceivable that, owing to some special circumstances of our occupation, our experience with 200 may induce a special intuitive knowledge of 200.) But numbers such as $\aleph_0$ and c are hardly to be attained in any such manner (by adding 1, that is).[20]

The number for the consummated infinity of natural numbers cannot be attained by the successive operation of adding 1. However astronomically vast a number may be, and however out of the question physically it may be to reach it by the successive addition of 1, a natural number can, in principle, be reached by the operation of addition. Aleph-null, however, cannot in principle be arrived at by starting with *any* natural number and continuing to add 1 to it. Furthermore, the cardinal number supposedly named by the symbol '$\aleph_0$' cannot in principle be conceived, or be an 'object of thought'. It is psychologically impossible to imagine a trillion farthings spread out before us, although we have some idea of what this would be like. But we have no idea of what it would be like to have in view an infinite number of farthings. This is because it is logically, not psychologically, impossible to view or imagine an infinite array of objects. The expression 'sees an infinite number of farthings arrayed before him' does not have a descriptive use in the language, unless the term 'infinite' is used in its popular meaning, of the huge.

The set of natural numbers 1, 2, 3, 4, ... is said to be countable, but it cannot be run through as it forms a nonterminating series. It has been argued that it is only physically impossible, not logically impossible, to run through the terms of an infinite series, because it is in principle possible to count off each successive number in half the time it takes to count off its predecessor: supposing 1 takes half a minute, 2 takes half of a half a minute, etc., then at the end of a minute *all* of the numbers of the infinite series will have been counted off, and the number of natural numbers, $\aleph_o$, will have been reached.[21] Without going into a detailed examination of this argument, it can be seen that since the natural numbers are endless, counting them off would be a task that could not come to an end − *after* which another task might be started. (A curious consequence of this argument is that if a minute were composed of an infinite geometric series of time intervals, $\frac{1}{2} + \frac{1}{4} + \frac{1}{8} + \ldots$, then since the series has no end, a minute could come to no end − which is a conclusion that only a metaphysician of the Eternal Now could accept.) Furthermore, the entire array of natural numbers cannot be given all at once (as an extension), for if it could be, then in principle it would be possible to run through the entire set of numbers. It is just as impossible to think of all the numbers *at once* as it is to finish running through them. If the series 1, 2, 3, 4, ... were a 'consummated' series of numbers, then it would be possible to run through them − as it is in theory possible to run through the series of numbers up to 73,583,197,773. And if they could be viewed all at once, as a whole, they could be run through.

It has been maintained that a collection the elements of which can neither be run through nor displayed all at once may nevertheless exist as a complete totality, and *in a sense* be given. According to this thesis, the succession of natural numbers covered by the expression '... etc. *ad infinitum*' form a completed set just as do the numbers up to the expression, and are given along with them but in a different manner. Thus, Russell has written:

... it is not essential to the existence of a collection, or even to knowledge and reasoning concerning it, that we should be able to pass its terms in review one by one. This may be seen in the case of finite collections; we can speak of 'mankind' or 'the human race', though many of the individuals in this collection are not personally known to us. We can do this because we know of characteristics which every individual has if he belongs to the collection, and not if he does not. And exactly the same happens in the case of infinite collections: they may be known by their characteristics although their terms cannot be enumerated. In this sense, an unending series may nevertheless form a whole, and there may be new terms beyond the whole of it.[22]

And also:

Classes which are infinite are given all at once by the defining property of their members.[23]

Aristotle held the theory of the concrete universal, according to which 'no universal exists apart from its individuals',[24] the Parmenidean implication being that whatever we can think of exists. But this is a philosophical view and has nothing to do with fact. It *is* an everyday fact, hardly worth mentioning, that we can and often do think of what does not exist: it is possible to entertain a concept to which nothing answers, e.g., the concept of a cat with five heads; and it is possible to entertain a defining property of a class which happens to be empty, e.g., the property of being a pterodactyl. A class that is determined by a given defining property ϕ may be empty, and entertaining a concept is no guarantee that there is anything answering to it. To say that a class is *given* by its defining property is to say only that its defining property is given; and to say that an infinite set is 'given all at once' by the defining property of its members is to say only that the defining property is given. The impression created is that more than this is being said, but this impression is delusive. Furthermore, to argue that an infinite collection can be known by its characteristics and that 'in this sense' an unending series may form a whole is merely to *assign* a sense to the phrase 'unending series which forms a whole': the phrase is arbitrarily made to mean the same as 'series which is known by its characteristic'. For example, saying that the series, 1, 4, 9, 16, ... forms a whole *in the sense* that its terms are characterized by being values of n^2 is to create by fiat a semantic identity: the phrase 'forms a consummated series' is *made* to mean the same as the phrase 'consists of successive values of n^2'.

There is a further consideration in support of the proposition that an infinity of elements can form an actual class. The idea that there cannot be a completed infinite series is connected with the idea that there cannot be anything beyond an infinite series. The idea behind this would seem to be that there can only be something after, or beyond, a series which forms a whole, and that since a series which comes to no end cannot form a whole series, nothing can come after it. Russell's answer to this is that 1 is beyond the infinite series $\frac{1}{2}, \frac{3}{4}, \frac{7}{8}, \frac{15}{16}, \ldots$, the implication being that an unending series can be consummated and form a whole class.[25] About the series of natural numbers he wrote:

... every number to which we are accustomed, except 0, has another immediately before it, from which it results by adding 1; but the first infinite number does not have this property. The numbers before it form an infinite series, containing ordinary finite numbers, having no maximum, no last finite number, after which one little step would plunge us into the infinite. If it is assumed that the first infinite number is reached by a succession of small steps, it is easy to show that it is self-contradictory. The first infinite number is, in fact, beyond the whole unending series of finite numbers.[26]

As in the case of the infinite series $\frac{1}{2}, \frac{3}{4}, \frac{7}{8}, \frac{15}{16}, \ldots$, with regard to which Russell states that 1 lies beyond the *whole* series, so in the case of the natural numbers he states that $\aleph_0$ lies beyond the *whole* series. Since the possibility of an unending series being a whole is at issue, Russell appears to be begging the question. Undoubtedly what he wished to say was that the fact that 1 lies beyond the series $\frac{1}{2}, \frac{3}{4}, \frac{7}{8}, \frac{15}{16}, \ldots$ shows that the series is a whole; and the fact that $\aleph_0$ lies beyond the series of natural numbers shows that $1, 2, 3, 4, \ldots$ is a whole, or a consummated series.

It is not clear what is meant by 'is beyond' or 'comes after' the whole of an infinite series. It is natural to say that 5 lies beyond $1, 2, 3, 4$; but 1 is not beyond the series $\frac{1}{2}, \frac{3}{4}, \frac{7}{8}, \frac{15}{16}, \ldots$ in this sense. The only sense in which 1 might be said to be beyond the series is that it is the *limit* of the series in the mathematical sense. We may be puzzled to know why the word 'beyond' rather than the more usual term is used, until we realize that 'beyond' suggests the idea of something finished which is followed by something else, i.e., the idea of one thing coming after another. The term 'limit' does not carry with it this suggestion, although there is a tendency sometimes to think of the limit of an infinite geometric series like $\frac{1}{2} + \frac{1}{4} + \frac{1}{8} + \frac{1}{16} + \ldots$ as being its arithmetical sum, which also suggests the idea of a completed series. But no series has an arithmetical sum whose terms cannot *in principle* be summed up; and this is not the case with regard to an unending series. Expressed somewhat differently, the series $\frac{1}{2} + \frac{1}{4} + \frac{1}{8} + \frac{1}{16} + \ldots$ has no sum because the sequence $\frac{1}{2}, \frac{3}{4}, \frac{7}{8}, \frac{15}{16}, \ldots$ has no last term; and this means that neither array of terms is a completed whole. Putting this aside, it can be seen that to contend against the view that 'there cannot be anything beyond the whole of an infinite series' by stating that '1 is beyond the whole of the infinite series $\frac{1}{2}, \frac{3}{4}, \frac{7}{8}, \frac{15}{16},$ ...'[27] is not to show that an infinite series is a whole, any more than to demonstrate that a given infinite series has a limit is to show that the series is a whole.

Similarly, to say that $\aleph_0$ is 'beyond the whole unending series of finite numbers' is not to show that the series exists as a completed whole. The series 1, 2, 3, 4, ... is said to increase without limit, but Russell states that $\aleph_0$ is the limit of the series. He writes:

The cardinal number $\aleph_0$ is the limit (in the order of magnitude) of the cardinal numbers 1, 2, 3 ... n, ..., although the numerical difference between $\aleph_0$ and a finite cardinal is constant and infinite; from a quantitative point of view, finite numbers get no nearer to $\aleph_0$ as they grow larger. What makes $\aleph_0$ the limit of the finite numbers is the fact that, in the series, it comes immediately after them, which is an *ordinal* fact, not a quantitative fact.[28]

One impression a close reading of this passage is liable to create is that a fast and loose game is being played with terminology: a series which increases without limit is said to have a limit, but in an *ordinal* sense; and $\aleph_0$ is said to come *immediately* after the series, although the series has no last term. The number 5 comes immediately after the series of numbers 1, 2, 3, 4. But if we stop to think about it, if we get behind the words, so to speak, we find that we have no idea of which it is for something to come *immediately* after a series which increases without limit. Light is thrown on the passage if we being it into connection with the assertion that '1 is beyond the whole of the infinite series $\frac{1}{2}, \frac{3}{4}, \frac{7}{8}, \frac{15}{16}, \ldots$'. What comes through is that the series 1, 2, 3, 4, 5, ... is represented as being like the geometric series:

$$\aleph_0 \text{ is the limit of (and is beyond)}$$
$$1, 2, 3, \ldots n, \ldots$$
$$1 \text{ is the limit of (and is beyond)}$$
$$\frac{1}{2}, \frac{3}{4}, \frac{7}{8}, \frac{15}{16}, \ldots \frac{2^n - 1}{2^n}, \ldots$$

To say that the series 1, 2, 3, 4, ... has a limit which lies beyond the whole of it, and that $\aleph_0$ comes immediately after it, induces one to think of the series as consummated. But this is not the same as showing that it is. We might say that this way of talking about the series shows nothing about it, but it does produce a change in the atmosphere.

The symbol '1' denotes a number which is the limit of the series $\frac{1}{2}, \frac{3}{4}, \frac{7}{8},$

$\frac{15}{16}$, . . . ; and the symbol '$\aleph_0$' is represented as denoting a cardinal number which is the limit of the series of natural numbers 1, 2, 3, 4, . . . The question is whether '$\aleph_0$' has in fact a use to denote a number, or whether we become dupes when with practice we reach a state of mind in which we think of '$\aleph_0$' as a numeral, like '15'. It has been seen that there is no way in which an infinite number of objects can be given. It is impossible to finish passing in review an infinity of entities, or to arrive at an infinite totality by counting. It is also impossible to envisage an infinity of objects spread out as a whole before us. The impossibility of counting the elements of an infinite set and of arriving at their number by adding them up is theoretical, as is also the impossibility of viewing an infinite array and noting its number, rather than entertaining the defining property of its members. The impossibility is logical, not one which is due to a psychological or a 'medical'[29] shortcoming. Any shortcoming which makes it impossible for us to carry out a task could in theory, if not in fact, be made good and the task brought within our reach. We know what it would be like to do things that are immeasurably beyond our actual abilities, like snuffing out the sun or jumping to Neptune, but we have no idea of what it would be to finish running through an infinite series or to see it as a completed totality. This is because the expressions 'comes to the end of an endless task' and 'sees an infinite totality before him' describe nothing whatever: the terms 'consummated infinite series' and 'actual infinite collection' have been given no application. Their actual use, regardless of the talk surrounding them, is neither to describe a series not to describe a collection. If the phrase 'the infinite series of natural numbers' described a consummated series, it would describe what in theory, if not in fact, we could run through. The phrase 'finished running through the series, 1, 2, 3, 4, . . . n . . .' would then have descriptive sense, which it does not.

There is no number which is the number of the totality of natural numbers, because there is no such totality. Russell asserted that 'It cannot be said to be certain that there are any infinite collections in the world'.[30] This observation carries with it the suggestion that the question whether there are infinite collections in the world is factual, to be investigated by empirical procedures. It should now be clear that the question is not a request for empirical information, and the statement that the series of natural numbers is not a totality of numbers is not an empirical statement. It follows that the statement that there is no number which is the number of the totality of natural numbers is not empirical: the phrase 'the number of the totality of natural numbers' does not describe or refer to a number. And the symbol '$\aleph_0$', which supposedly denotes the number referred to by the descriptive phrase, is, unlike the numeral '15', not only not an 'ordinary' numeral, it is not the name of a number at all.

It need hardly be pointed out that we do speak of the existence of an infinite number of natural numbers and of the existence of an infinite number of rational numbers; and it would be foolish to deny that sentences declaring the existence of an infinite number of terms and the existence of an infinite series, etc., are perfectly intelligible. How is this fact to be brought into line with what has just been said about '$\aleph_0$', which is represented as denoting the number of their terms? Consider the proposition that there exists an infinite number of prime numbers. Proving it is the same as proving that the hypothesis that there is a greatest prime number is self-contradictory: supposing P to be the greatest prime, then $P! + 1$ either is itself prime or contains a prime factor which is greater than P. That is, the proof of the proposition that there is an infinite number of primes consists of showing that the concept of a greatest prime implies a contradiction, and thus that no number answers to it – just as no number answers to the concept of an integer between 5 and 6. This means that any expression whose meaning is the concept (in the English language the expression 'greatest prime number') has no use to describe a number. Restated in terms of language, demonstrating the proposition expressed by the sentence 'There is an infinite number of primes' is nothing in addition to demonstrating that the proposition expressed by the sentence 'There is a greatest prime' is self-contradictory, and this in turn is the same as showing that the phrase 'greatest prime number' has no use to describe or to refer to a number. The important point to grasp is that a sentence which declares the existence of an *infinite number* of terms in a mathematical series means nothing different from a sentence which declares the *nonexistence* of a term answering to a putative description, 'the last number' or 'the least number' or 'the greatest number', and the like. Undoubtedly it was a kind of recognition of this fact that was responsible for John Locke's observation that we have no *positive idea* of the infinite.[31] The sentence 'There exists an infinite number of natural numbers' says the same thing as does the sentence 'There is no greatest natural number'. And the sentence 'There is no greatest number' conveys, without *expressing* what it conveys, the verbal fact that the expression 'greatest natural number' has no use to refer to a number. In other words, the *implicit* import of the sentence 'There exists an infinite number of natural numbers' is wholly verbal and negative, to the effect that a certain expression has no application in the language of mathematics.

It can now be seen why it implies no contradiction to say that an infinite number is not a number, and we can understand why 'when infinite numbers are first introduced to people, they are apt to refuse them the name of numbers'. If we give a moment's thought to Locke's remark that '. . there is nothing yet more evident than the absurdity of the actual idea of an infinite number'[32] and consult the workings of the language of the infinite, we will see that talk of the consummated infinite is bogus. Wittgenstein made the remark that what the bedmaker says is all right, but what the philosopher says is all wrong, and we might now be inclined to think that what the mathematician shows about infinite series, denumerable, nondenumerable, etc., is all right, but that what the philosophical mathematician says about the actual infinite is all wrong. Instead of declaring that what the philosophical mathematician says is wrong, that he is making mistakes, it is more enlightening to think of him as making up a special language-game, in which '$\aleph_0$' and other symbols, e.g., 'c', are treated *as if* they are the names of numbers, for the special aura doing this provides for a certain part of mathematics. It is a secure maxim that philosophers do not make mistakes, mistakes to which they are incorrigibly attached. Instead, they play games with language for whatever subterranean value they may have. A mathematician who surrounds his work with a dramatic language-game undoubtedly derives hidden satisfaction from it.[33]

The assertion that an unending series forms a whole, or a completed extension, 'in the sense' that the characteristic of the series is known is an imaginative way of speaking about a rule (whether explicitly formulated or not) for generating terms in serial order, no term being the last that is constructible by the rule. This is what is meant and all that is meant by saying that a series is infinite. The *substance* behind talk of the completed infinite, the actual mathematics behind it, is just the explication of the characteristics of formulas for constructing series. To revert to Russell's talk about finite and infinite numbers, the terms 'finite' and 'infinite' have actual applications only to series, not to numbers: the number 7 is not a 'finite' number, it is a number. A series is said to be finite when a number applies to the set of its terms, as in the case of an arithmetic series, and it is said to be infinite when it is linked with a rule which implies the logical (not physical or psychological) impossibility of any term being the last in the series constructible by it. In its actual use 'infinite number of terms' does not refer to a *number* of terms. And to say that a series contains an infinite number of terms is not to make a statement about the number of terms in the series: 'series which contains an infinite number of terms' means the same as 'infinite series', which in turn means the same as 'series that is generated by a formula with regard to which it is *senseless* to say that it generates a last term'. The likeness between such expressions as 'the number 37', 'huge number', and the term 'infinite number' lies in their grammar, to use a word made popular by Wittgenstein: like them it is, grammatically, a substantive expression, but unlike them it neither describes nor names a number.

The grammatical likeness between the term 'infinite number' and terms like 'large number' and 'small number', as well as the grammatical likeness between 'infinite series' and 'finite series', creates possibilities of playing exciting games with language. The idea that this is what mathematical philosophers of the infinite are doing helps us understand the import of the remark that with practice '$\aleph_0$' will acquire 'the same significance for us as the number 15'. The underlying meaning plainly is that in time people will enter into the language-game and will come to *feel* about '$\aleph_0$' much as they feel about numerals, especially numerals which denote prodigious numbers. The idea that philosophical mathematicians who are concerned to provide a 'theoretical' background for their actual mathematics, who, in other words, wish to have one or another philosophical 'centre piece'[34] for mathematics, gives us insight into the mysterious and continuing dispute between finitists and Cantorians. No mathematical statement or demonstration is actually in dispute. And it gives us an improved understanding of the divergence of opinions represented by Galileo, Leibniz, and Cantor.

To go to the matter directly, a philosophical mathematician like Leibniz (and a person to whom 'infinite numbers are first introduced') seems to be impressed by the semantic dissimilarity, i.e., the difference in use, between 'infinite number', 'huge number', and like expressions, and unimpressed by the similarity of their grammar. In his opinion, it would seem, the grammatical similarity tends to cover up an important difference in the actual use of terminology. And in play, if not in fact, he 'corrects' this shortcoming in language. He declares infinite numbers to be self-contradictory, which, like the greatest prime and a rational number whose square is equal to 2, do not exist. His argument is that no whole number can be equal to a fraction of itself, and since a whole infinite number would be equal to a fraction of itself, e.g., $2\aleph_0 = \aleph_0$, it is impossible for there to be an infinite number.

Galileo, it would seem, was intrigued by the grammar of the term 'infinite number' and came out in favor of treating the term *as if* its use was to refer to a number. Equally with Leibniz, he was aware of the fact that as John Locke put it 'nothing is plainer than the absurdity of the actual idea of an infinite number', which is a nonverbal way of stating the verbal fact that 'infinite number', like 'last natural number' and unlike 'the smallest prime number greater than 1,000,000', does not function in the language to refer to or to describe a number. Nevertheless he chose to group it, artificially, without changing its actual use, with what might be called substantive number expressions. Parenthetically, it is worth remarking that although Leibniz and Galileo give opposing answers to the question, 'Is an infinite number a number?' their answers are not the result of any difference in their knowledge of numbers, which would be inexplicable if the question were a request for information about them. To put it in John Wisdom's way, the question is not a request for mathematical information; it is a request for a redecision with regard to the term 'infinite number', as to whether to classify it with substantive number expressions. The answers represent opposing linguistic decisions, which make no difference to the doing of mathematics and thus can be argued interminably. Treating 'infinite number' as if it is a substantive number expression which gives the number of terms in the series 1, 2, 3, 4, . . . generates the paradoxical property of an infinite set that some only out of all of its elements are no fewer than all of the elements. Leibniz claimed that the property was self-contradictory, and thus that the idea of an infinite number was self-contradictory. Galileo took a different view of what the paradoxical property showed about infinite numbers: its possession is *proof* that '.. the Attributes or Terms of Equality, Majority, and Minority, have no place in Infinities, but are confin'd to terminate quantities'. What, according to Leibniz, demonstrates the impossibility of infinite numbers according to Galileo demonstrates that, unlike 7 and 23, infinite numbers are not comparable: the number denoted by '$\aleph_0^2$' cannot be said to be either less than or equal to the number denoted by '$\aleph_0$'. What can be seen here are not different opinions regarding what is *entailed* by the possession of a certain property, but different ways of marking the unlikeness between 'infinite number' and substantive number expressions. Galileo classifies the term with substantive number expressions, and marks the difference between them by stating that infinite numbers are mysterious, elusive numbers which cannot be compared with each other, numbers over which our 'finite understanding' creates difficulties. No actual entailment-claim is in question. Only a way of marking the difference between the actual use of terminology is being put forward.

Cantor, who according to Russell transformed a property formerly thought to be self-contradictory into a 'harmless definition' of infinity, goes against both Leibniz and Galileo. In his view, infinite numbers are not self-contradictory nor are they mysteriously different from 7 and 23 and $3^3 - 2^2$ in not being comparable in terms of less than, equal to, and greater than. Two sets are said to have the same cardinal number (Cantor's term was '*Mächtigkeit*'), or to be equal, if there exists a (1-1)-correspondence between their elements. In the words of E.T. Bell, 'Two sets are said to have the same *cardinal number* when all the things in the sets can be *paired off* one-to-one. After the pairing there are to be no unpaired things in either set'.[35] He states that 'Cantor proved that the set of all rational numbers contains precisely as many members as the (infinitely more inclusive) set of *all* algebraic numbers'.[36] He might also have said that Galileo *proved* that the set of all rational numbers contains precisely as many members as the infinitely less inclusive set of all the squares of the rational numbers. There is no reason for thinking that Galileo would have agreed about what he had *proved*. Russell stated that the property of having no more terms than does a proper subset of itself shows a set to be infinite, and that it shows that the number of natural numbers is infinite. There is no reason for thinking that Russell's words would have made Leibniz admit to being mistaken. This is because there is no true opinion and no false opinion about whether infinite sets exist as wholes and about whether the conceptions of *equal to, less than*, and *greater than* apply to them. What can be seen is that 'transforming' a supposedly self-contradictory property into a 'harmless definition' of the term 'infinite number' comes down to marking the difference between the use of 'infinite number' and that of substantive number expressions. Like Galileo, Cantor classifies 'infinite number' with substantive expressions like 'large number', but instead of marking the unlikeness between them as Galileo does, he marks it differently. Galileo's conclusion that infinities are not comparable (which is a hidden way of stating that 'infinite number' does not refer to a number) results in one kind of mystification. The Cantorian conclusion that a fraction of an infinite number can be equal to the whole number (which is also a hidden way of stating that 'infinite number' is not a substantive number expression) results in another kind of mystification. Nevertheless, the claim that the terms 'equal to', 'less than', and 'greater than' apply not only to the natural numbers, but also to infinite numbers, brings the term 'infinite number' into line with substantive number expressions – which makes it possible with practice to come to think of '$\aleph_0$' as having a use like that of '15'.

In the theory of infinite sets some sets are said to be equal to each other or to have the same cardinal number, namely those whose elements are (1-1)-correlatable, as are, for example, the terms of the sets of natural numbers, their squares, and the squares of their squares:

$$1, 2, 3, 4, \ldots \ldots$$
$$1^2, 2^2, 3^2, 4^2, \ldots \ldots$$
$$1^{2^2}, 2^{2^2}, 3^{2^2}, 4^{2^2}, \ldots \ldots$$

Some infinite sets are said not to be equal to each other and thus to have different cardinal numbers. Cantor showed that the real numbers (roughly, numbers which can be represented by unending decimals) are not (1-1)-correlatable with the natural numbers, the conclusion being that c, the number of the totality of the real numbers, is not equal to $\aleph_0$. Not to go into the actual demonstration, he showed that assuming the totality of real numbers to be in an array, it is possible to produce real numbers which are not in the array. Hence, there can be no (1-1)-matching of the real numbers with the natural numbers. Imitating the language of E.T. Bell, *after* the pairings of the reals with the natural numbers there will be unpaired terms left in the set of cardinal number c. The fact that the natural numbers cannot be matched (1-1) with the real numbers is taken to imply that there are *more* real numbers than there are natural numbers, and thus that c is a *greater* cardinal number than $\aleph_0$, just as 9 is greater than 7. The reason is that since c is not equal to $\aleph_0$, it must be greater $- c > \aleph_0 -$ in contrast to $\aleph_0^2 = \aleph_0$. The assimilation of transfinite cardinal arithmetic to the natural number arithmetic, which is characterized by the concepts *equal to, greater than*, and *less than*, is impressive. The infinite and the super-infinite to all appearances are tamed to the harness of the 'finite' numbers.

The theory of the actual infinite creates the 'crude miracle' of a proper part of a set being no less than the whole set, and to this it adds the further miracle of an infinity that is greater than other infinities, an inexhaustible that is more inexhaustible than in infinitely inexhaustible. Bell writes: '... try to imagine the set of *all* positive rational integers 1, 2, 3, ..., and ask yourself whether, with Cantor, you can hold this totality – which is a 'class' – in your mind as a definite object of thought, as easily apprehended as the class *x, y, z* of three letters. Cantor requires us to do just this thing in order to reach the *transfinite* numbers which he created?'[37] It will be plain that just as we cannot hold the set of all positive rational integers in our mind as an object of thought (not only not as easily as we can hold a set of three elements but because it is logically out of the question), so we cannot envisage the set of real numbers as a whole. Just as we have no idea of what it would be to apprehend *all* the terms of an unending set, so we have no idea of what it would be to entertain all the terms of a set that has more terms than an unending set. The transfinite numbers which Cantor 'created' are creations of a different kind from what it is natural to take them to be: '$\aleph_0$' and 'c' are *als ob* names of numbers.

There can be no doubt that '$\aleph_0$' and 'c' do have a use in mathematics, although their use, except in semantic appearance, is not to refer to numbers. What their actual use is, as against their apparent use, can now be seen: '$\aleph_0$' refers to rules or formulas for constructing series of terms, no term of which is the last constructible by the formula, and 'c' refers to rules for constructing from sets of terms new terms which are not in the original sets, however large those sets are made. There is no hint of a miracle or of paradox in the fact that for every term constructible by n^2 a uniquely related term is constructible by n^{2^2}, but an exciting paradox makes its appearance when we state that there are just as many natural numbers of the form n^2 as there are natural numbers of the form n^{2^2}, or that some only out of all the members of a certain collection are no fewer in number than all of the members. It is interesting, but nothing strange, to be shown, by the so-called diagonal method, that from any array of real numbers a new real number can be constructed which is not in the array, and thus that terms constructed by n^2

cannot be exactly matched with those constructed by the diagonal rule. But there is excitement and strangeness in being told that there is an infinity which is greater than the infinity of natural numbers. The excitement and the appearance of the miraculous are produced not by what is said but by the way it is said. Stating that the real numbers are not (1-1)-correlatable with the natural numbers stirs up no thoughts of the miraculous, but saying that the set of real numbers is more huge than the infinite set of natural numbers creates in some people awe and wonder, although what is being said is the same. One way of speaking opens the gate to paradise for some mathematicians, although to other mathematicians it looks like a disease-ridden land. The actual mathematics is the same for all. It is the surrounding philosophical talk, the mathematical theatre, which attracts some and repels others; but the philosophical talk has no effect on the actual mathematics in transfinite number theory.

NOTES

[1] Cited by P.E.B. Jourdain, *The Philosophy of Mr. B*RTR*ND R*SS*LL*, p. 63.
[2] *Our Knowledge of the External World*, pp. 160–61.
[3] Ibid., p. 170.
[4] *The Blue Book*, pp. 45–6.
[5] *Philosophical Investigations*, p. 48.
[6] *Our Knowledge of the External World*, p. 199.
[7] My italics.
[8] Bertrand Russell, op. cit., p. 173.
[9] Ibid., p. 181.
[10] *Introduction to Mathematical Philosophy*, pp. 78–9.
[11] Ibid., p. 79.
[12] E.T. Bell, *Men of Mathematics*, p. 567.
[13] Leibniz, *Philosophische Schriften*, Gerhardt's edition, Vol. I, p. 338.
[14] *Our Knowledge of the External World*, p. 194.
[15] 'Mathematics and the Mathematicians', in *Mysticism and Logic*, p. 86.
[16] *Our Knowledge of the External World*, p. 182.
[17] *Introduction to the Foundations of Mathematics*, p. 87.
[18] Ibid., p. 81.
[19] Hans Hahn, 'Infinity', in James R. Newman's *The World of Mathematics*, Vol. 3, p. 1602.
[20] Op. cit., p. 101.
[21] See Russell, 'The Limits of Empiricism', *Proc. of the Aristotelian Society* 36. This paper is a critique of Alice Ambrose's papers on finitism in mathematics. For her reply, see her *Essays in Analysis*, Chapter 4.

[22] *Our Knowledge of the External World*, pp. 181–2.
[23] Ibid., p. 156.
[24] *Metaphysics*, Book Z, translated by W.D. Ross.
[25] Op.cit., p. 173.
[26] Op. cit., p. 181.
[27] Op. cit., p. 173.
[28] *Introduction to Mathematical Philosophy*, p. 97.
[29] Bertrand Russell, 'The Limits of Empiricism', op. cit., p. 143.
[30] *Introduction to Mathematical Philosophy*, p. 77.
[31] See *An Essay Concerning Human Understanding*, Book II, Ch. 17, Sec. 13.
[32] Ibid., Sec. 8.
[33] I may allow myself a conjecture as to one component of the hidden satisfaction. Maimonides said that to study nature is to study God, and this certainly is implied in Spinoza's philosophy. The idea that cannot fail to cross one's mind in connection with Cantor is that to study mathematics is also to study God. It should be remembered that Kronecker declared that God made the integers. It is hard to think that the idea of the consummated infinite is not in some way unconsciously linked with the idea of God as the consummation of infinite greatness.
[34] Taken from Freud's well-known observation, 'Putnam's philosophy is like a beautiful table centre; everyone admires it but nobody touches it' (in Ernest Jones, *Free Associations. Memories of a Psycho-analyst*, p. 189). Wittgenstein observed that the philosopher's labor 'is, as it were, an idleness in mathematics' (in *Remarks on the Foundations of Mathematics*, p. 157).
[35] *Men of Mathematics,* p. 566.
[36] Ibid., p. 565.
[37] Op. cit., p. 567.

CHAPTER IX

THE PASSING OF AN ILLUSION

*Once I had been able to tear aside an illusion that had
previously dimmed my vision. . . the insight thus gained
was never lost.*

Ernest Jones, *Free Associations.*

A belief that derives its strength from a wish rather than from evidence is an
illusion. It sometimes happens that a belief which springs from and is sustained
by a wish later turns out to be true, but when this happens the belief does not
lose its quality of having been an illusion. The world is indifferent to our
wishes, and illusions seldom are realized; the old saying that if wishes were
horses beggars would ride certainly applies to them. As is well known, Freud
declared religion to be an illusion. To convince oneself of the correctness of
his claim no more need be done than to notice the difference between the
demands for evidence a scientist who is also religious places on his scientific
propositions and those he places on the propositions laid down for him by his
religion. The thesis developed in these pages is that academic, reasoned philosophy, exemplified by the central doctrines of Aristotle, Anselm, Descartes,
Hume, G.E. Moore, is an illusion, but an illusion which is different in a
remarkable way from that presented by religion. To put it roughly, a set of
religious beliefs is about reality and what we can expect from it. A philosophical system, which appears to be a set of propositions about the world,
is not at all what it appears to be. Language can be used to express illusions;
but things can be done with language to create the deceptive illusion that
words are being used to express propositions about things. Philosophy, it
turns out, is a linguistically contrived illusion. It is hardly an exaggeration to
describe it as an unconscious semantic swindle, by which the swindler is also
swindled. If one becomes attentive to a philosopher's defense of his discipline,
what stands out it its likeness to the defenses of religion, underlying which is
a seriously weakened ability to be critical. Thus, a philosopher whose attention is called to the bewildering fact that philosophy, which is by no means
a young discipline, has not achieved a single uncontested result will show a
remarkably weakened sense of curosity, coupled with blithe unconcern.

Technical philosophy, which supports its claims with reasoned arguments, holds an ambiguous place amongst the disciplines that profess to give us knowledge of things. It is taken by philosophers, and by many people who have peripheral acquaintance with it, to be a kind of science, but not just one amongst the sciences. It is thought to be more basic than the standard sciences. Not only does it make deeper investigations into the nature of things, it also critically examines the presuppositions on which the special sciences rest. It is also realized, sometimes explicitly, that philosophy makes no use of the modes of verification employed in astronomy, chemistry, biology, or any of the other natural sciences. It has no laboratories of its own: there is no such thing as a philosophical experiment. And if, as the language of philosophers sometimes suggests, it resorts to observations, it seems to have no need for instruments which extend the reach of our senses. Some philosophers have the notion that an underlying wish of many theoretical physicists is the desire to attain a philosophical understanding of 'the inner nature of reality'.[1] Nevertheless, it is realized by everyone, however indistinctly, that a physicist who embarks on a *philosophical* investigation into the nature of things — the outcome of which are such views as that things are collections of minds and that there are abstract entities — leaves behind him the observational and experimental methods of the sciences. It would seem that a philosophical investigation of things, which probes into their ultimate nature, requires methods more refined than the gross ones of the special sciences, methods which differ from them *in kind*. But what these methods are is shrouded in mystery. One philosopher said: 'Whistling in the dark is not the method of true philosophy'.[2] These words have the ring of a magical reassurance formula. It may well be that a philosopher who thinks 'that philosophical problems are genuine and that they are capable of being solved'[3] is whistling in the dark, and is trying to reassure himself against the failure of philosophy to produce results.

A number of the great figures in philosophy who became sensitive to the total absence of undisputed philosophical propositions have placed the blame for this malady on the absence of a correct method of procedure, and have tried to make good the lack. Names that will immediately come to everyone's mind are Descartes, Kant, Bertrand Russell, G.E. Moore. As was to be expected, the riot of claims and counter-claims has continued with abandon and unabated enthusiasm. But there is not the slightest sign that reasoned philosophy will eventually reach a stage when it can begin to boast of ordered progress. Nevertheless, the work of Descartes (whose model of an ordered science was mathematics) and that of other philosophers has tended to bring philosophy out of its obscurity and to make somewhat more visible its inner character.

The picture of the philosopher as the investigator of reality who goes about his work without any of the tools of science demands an explanation. We should not treat anyone with respect who tells us what is in books which he has never opened and into whose pages he has never looked. But the philosopher is, on the surface at least, treated with respect, even though one sometimes gets the impression that it is the word 'philosophy' rather than its special subject matter that is deferred to. Freud has remarked that philosophy 'is of interest to only a small number even of the top layer of intellectuals and is scarcely intelligible to anyone else'.[4] The philosopher would lose the interest of even the top layer of intellectuals, however, if he could produce no credentials attesting to the validity of his pursuit of cosmic knowledge, no methodology which would support the notion that his pursuit was both sober and responsible. C.D. Broad, who explicitly called attention to the great difference between the method of philosophy and that of the natural sciences, stated: 'Experiments are not made, because they would be utterly useless.'[5] Philosophy resembles pure mathematics 'at least in the respect that neither has any use for experiment'.[5] Broad distinguishes between speculative philosophy and what he calls critical philosophy, which is 'the analysis and definition of our fundamental concepts, and the clear statement and resolute criticism of our fundamental beliefs'.[6] The criticism of a belief is not different from the analysis of a concept or a proposition, and the claim about the method employed by critical philosophy that comes out distinctly is that it is the analytical scrutiny of concepts. Broad does not say what the method is for checking the statements of speculative philosophy, whether it is analysis or experiment, but he does say that critical philosophy performs a task not performed by any of the natural sciences: 'The other sciences *use* the concepts and *assume* the beliefs: Critical Philosophy tries to analyze the former and to criticize the latter. Thus, so long as science and Critical Philosophy keep to their own spheres, there is no possibility of conflict between them, since their subject matter is quite different.'[7] No one can fail to see the striking similarity between Broad's attempt to reconcile critical philosophy with the natural sciences and the attempt to reconcile religion with them.

What the analysis of concepts (and of propositions) is, and what it is supposed to achieve for us is obscure. Some philosophers think that it consists in making explicit the components of the concepts we have of things, that doing this consists in making 'our ideas clear', and thus that it can yield no new information about the nature of the things falling under the concepts. Wittgenstein expressed this notion in the following way: 'A philosophical work consists essentially of elucidations. Philosophy does not result in 'philosophical propositions', but rather in the clarification of propositions.'[8] Other philosophers, however, have the idea that the analysis of a concept is capable of yielding new information about the things exemplifying it, that analysis is capable of yielding facts about the inner consistution of things. The underlying suggestion is that analysis is a more refined method of investigation than either experiment or observation, and that it continues at the point where these must leave off. Still other philosophers think that what is called the analysis of concepts is just the analysis of the uses of terminology in a language. This idea represents philosophy as being a kind of linguistic investigation. The picture of the philosopher as the investigator of language seems less mystifying that the picture of the philosopher as probing into concepts, but each presents us with an enigma.

On the view that philosophical analysis is just clarification, whether it articulates concepts or explicates the rules for the use of terms in a language, philosophy cannot be construed as an investigation of the world, an investigation which informs us of the existence of things and of how they operate with respect to each other. Analysis may be preliminary to an ontological investigation but cannot be such an investigation itself. And plainly, if philosophical methodology is confined to analysis, philosophy can tell us nothing about the world and can give the philosopher nothing that he seeks. On the view that the analysis of our concepts is an instrument for obtaining knowledge of the world we find ourselves in the presence of a riddle. For this view implies that the concept of a thing contains hidden information about the nature of the thing. To see this it need only be realized that the meaning of a word, for example 'planet', is a concept: to say that a general word names a concept is just another way of saying it has a meaning. Thus the analysis of a concept is just the analysis of the meaning of a word. Now if the analysis of the meaning of a word is to give us new information about things, we shall have to suppose either that we have given more meaning to our words than we are aware of or that our words have more meaning than we have assigned to them. Neither alternative is acceptable. Words do not have meanings by nature, meanings they have not been given; and we can hardly suppose that we give to words more meanings than we consciously assign to them. But in either case analysis of the meaning of a word would be of no use. The analysis of a known meaning of a word cannot bring to light an unknown meaning, since the unknown meaning cannot be *part* of the known meaning. It would seem that no inference from concepts to things can be the result of an analysis.

Kant said that we find in nature what we ourselves have put there. Reformulated in terms of concepts, this comes to saying that we find in the meanings of our words what we ourselves have put into them. But sober reflection should make us see that no philosopher actually thinks that we put hidden information about things into the meanings of our words, something we unwittingly do and later *discover*. This idea supposes that we have more knowledge of the nature of things than we actually have, knowledge that we had prior to the investigation of our meanings. The underlying fantasy is that to which Plato gave explicit expression in the *Phaedo*, i.e., his reminiscence theory, the theory that we are born with unremembered knowledge. The other view, viz., that philosophical analysis is just the analysis of the use of terminology, implies an odd delusion which we cannot really attribute to any philosopher. This is that unknown facts of ontology can be inferred from a scrutiny of words. In sum, the inferential leap from concepts to things, as well as the leap from words to things, cannot be justified by the analysis of either words or concepts. And how anyone could *appear* to think otherwise will have to be explained.

The picture of the philosophical cosmologist is that of someone who obtains knowledge of things by looking more deeply into usage than the lexicographer or by reflecting in a special way on concepts. The reason given in support of a philosophical claim, e.g., that motion does not exist, or that in addition to the concrete things of sense there are abstract entities, is not a piece of empirical evidence, either observational or experimental. To some philosophers it appears to be the eliciting of the consequences of a concept, while to others it appears to be the explication of the rules for the use of expressions. In either case the philosopher is pictured as trying to obtain knowledge of what there is at a remove from what there is. And to many people the attempt to do this suggests that the philosopher is suffering from some sort of delusion. Thus, to C.D. Broad the linguistic philosopher (who behaves is if a philosophical examination of usage will teach him things about the world) seems to be suffering from 'one of the strangest delusions that has ever flourished in academic circles'.[9] But the same words apply to the philosopher who imagines that analytical examination of concepts will give him new information about things. To suppose that factual information about things is obtainable by studying objects other than the things themselves is to suffer from a bizarre delusion.

The philosopher appears to suffer from this delusion while also being free from it. His talk throws an aura of unreality over his work, while his behavior is that of the normal man who knows how to live in the world. G.E. Moore has remarked on a paradox which deserves more attention than it has received, namely, that philosophers '. . . have been able to hold sincerely, as part of their philosphical creed, propositions inconsistent with what they themselves *knew* to be true.'[10] And if, as seems to be the case, a philosopher consults either usage or concepts rather than things in the attempt to discover facts about them, basic or otherwise, Moore's paradox clearly applies to him: philosophers have been able to believe that they can discover facts of ontology by consulting verbal usage or concepts while knowing their belief to be false. To help explain this odd state of affairs, Moore observed that a philosopher is able to hold his unusual beliefs in what he called 'a philosophic moment'. But whatever the bewitchment is that takes over in a philosophic moment, the philosopher's behavior remains normal. He does not give the impression of being a somnambulist or in any way suffering from an abnormal state of mind.

Philosophical analysis is represented as an ultra-refined scientific instrument which permits the philosopher to determine the nature of structure of reality without leaving his study. But it would indeed take remarkable powers of self-deception to be able to accept this representation at face value. We have to say that he carries on his work seriously, which certainly implies self-deception of some sort; but we also have to say that *in some way* he is not deceived. To illustrate briefly, a philosopher who seriously says that motion does not exist is in some way deceived and yet is not deceived, as is shown by the rest of his talk and his behavior. It will be clear that philosophy is in an equivocal position, which would explain its being approached with ambivalence. Philosophy uses reason, which wins our respect, but attempts to obtain

knowledge of things without going to the things, which makes it seem ludi-
crous. Freud exemplified the ambivalence many people feel toward philosophy,
as the following passage shows.

Philosophy is not opposed to science, it behaves like a science and works in part by the
same methods; it departs from it, however, by clinging to the illusion of being able to
present a picture of the universe which is without gaps and is coherent, though one
which is bound to collapse with every fresh advance in our knowledge. It goes astray in
its method of over-estimating the epistemological value of our logical operations and by
accepting other sources of knowledge such as intuition. And it often seems that the
poet's derisive comment is not unjustified when he says of the philosopher:
> Mit seinen Nachtmützen und Schlafrockfetzen
> Stopft er die Lücken des Weltenbaus[11].[12]

It is clear from this passage that Freud had two conflicting ideas about
philosophy: it works in part at least by the methods of science but it nevertheless is the target of jibes. The background picture Freud seems to have had
of the philosopher is that he is a kind of scientist who cuts a comical figure.
Ernest Jones, who interested himself in philosophy, also had conflicting
ideas about it. Sometimes it seemed to him to consist of 'tenuous sophistries
that have no real meaning',[13] and at other times it seems to be a highly
abstract and important discipline. Thus, he criticized Jung who postulated a
psychic toxin which poisons the brain by remarking that 'his grasp of philosophical principles was so insecure that it was little wonder that they later
degenerated into mystical obscurantism'.[14] About Greek science, however, he
said: 'It is no chance that when the Greek genius faltered at the threshold of
scientific thought by disdaining the experimental method and enmeshing
itself in the quandaries of philosophy, it was medical study alone which
forced it into some relationship, however strained, with reality.'[15] There can
be no question that some perception into the workings of philosophy created
in Jones conflicting attitudes to it, which he was able temporarily to resolve
only by attributing great mental power to philosophers and disparaging his
own ability in 'abstract fields'.[16]

The reasons Freud gave for the condition philosophy finds itself in are that
it overestimates the 'epistemological value of our logical operations' and that
it makes use of intuition. It is not clear what Freud had in mind by intuition,
nor is it clear what he meant by logical operations. He seems to be alluding to
an idea entertained by many philosophers from Leibniz on, the idea expressed
by Bertrand Russell's statement that logic is the essence of philosophy. But

the idea that logic is the tool of philosophical investigation is nothing different from the idea, going all the way back to Parmenides, that analysis is capable of uncovering basic truths about reality. As has already been remarked, philosophical analysis is a method which a philosopher can only use at a remove from the phenomena about which he professes to be seeking knowledge. His placing exclusive reliance on philosophical analysis is bound up with a 'disdain' for the experimental method. It succeeds only, to use Jones' words, in enmeshing him in the quandaries of philosophy, or as Wittgenstein graphically put it, in imprisoning the fly in the fly-bottle. If we look again at the passage quoted from Freud, what suggests itself is that the philosopher works under the domination of fantasied omniscience; which is concealed by talk of logic and analysis. Freud's remark that the philosopher overestimates the epistemological power of logic is a distorted way of saying that the philosopher imagines that he can plumb the secrets of nature by the powers of his mind alone. This of course fits in with what *in some way* is known to everyone, that the philosopher does not trouble to go to nature herself in his quest for knowledge of the essential nature of things.

The philosopher indeed whistles in the dark, for fantasied omnipotence of thought is the mainspring of philosophical methodology. Spinoza, who in his *Ethics* lays out the geography of the cosmos, declared that the order and connection of things is the same as the order and connection of ideas. Parmenides held that thought coincides precisely with being, and Hegel laid down as his basic postulate that the real is the rational. We now have two pictures of the philosopher. One is the familiar picture of the philosopher as the cosmic explorer whose method does not require him to leave his study. The philosopher who naturally comes to mind in this connection is Spinoza: he did not find it necessary to leave his room in The Hague to delineate reality. The second picture, which is behind the first, is that of someone who has not been able to relinquish his narcissistic attachment to the power of his own thought, in other words, of someone who is still under the domination of mental megalomania. Aborigines speak of time before memory as dream time, and it would seem that a philosopher has found a way of giving satisfaction to a stage of development in his own dream time in the practice of philosophy. Behind the facade provided by philosophical analysis he is able to give gratification to the wish to encompass reality, like Divinity, with his mind alone. His wish keeps him at a sufficient distance from the activity to prevent him from seeing what it is, and his activity gives him imagined realization of his wish. Technical philosophy may truly be described as a sophisticated but delusive way of satisfying a wish which cannot be satisfied in reality. G.E. Moore invoked the expression 'philosophic moment' to minimize the strange state of mind that permits a philosopher to say without embarrassment such things as that time is unreal and that things do not exist. We have a possible explanation, at least in part, of what a philosophic moment is, namely, that for the moment he is under the domination of a unconscious wish. This, however, cannot be the entire account. There needs to be a further explanation of philosophical activity which will help us understand why philosophical statements that appear to be about the world can be investigated away from the world.

The narcissism which finds expression is an unconscious belief that one is in possession of stupendous mental powers acts as a sentinel against trespassers who might be led by curiosity to try to obtain a closer view of things. The philosophical investigation of existence can be nothing more than a delusive way of gratifying the wish for omniscience, and the end results of the investigation as well as the investigation itself are protected from our scrutiny by the strength of the narcissistic investment in them. Without protection the philosophical creation would evaporate into thin air and leave behind a semantic residue that could deceive no one. If we can ward off the influence of a philosophic moment and draw sufficiently close to a philosophical theory, e.g., the theory that matter does not exist, what we shall see is a contrived re-edition of familiar terminology the sole function of which, apart from its connection with unconscious material, is to create an intellectual illusion, and not to describe or declare the existence of a kind of object.

To get a preliminary understanding of what goes on in philosophy it is necessary to see clearly the difference between using an expression in a statement about a thing and mentioning that expression in a statement about its use in the language. This distinction is elementary but important to keep in mind. For it is possible to formulate sentences which make no explicit mention of terminology but whose sole content is, nevertheless, verbal; and therein lie the seeds of philosophy. To anticipate, it is by means of the non-verbal facade. i.e., the ontological form of speech, that the philosopher, whose work consists of nothing more than verbal manoeuvrings, is able to create the image of himself as a cosmological cartographer.

The sentence 'A cockerel is incapable of doing arithmetic' uses the word 'cockerel' to refer to an object about which it makes as assertion. The sentence ' "Cockerel" means young male fowl' mentions the word 'cockerel' instead of using it. The first sentence may be said to have ontological import: it uses language to state a claim that is not about the use a term has in a language. The second sentence states a claim, true or false, about the use, or the literal meaning, of a term, and clearly has only verbal import. The sentence 'A cockerel is a young male fowl' is unlike the verbal sentence ' "Cockerel" means young male fowl' in that it does not mention the word 'cockerel', and in this respect it is like the nonverbal sentence 'A cockerel is incapable of doing arithmetic'. The content of the sentence 'A cockerel is a young male fowl' is, nevertheless, verbal, and this tends to be veiled by the nonverbal form of speech in which the sentence is formulated. Understanding it, just as understanding the ontological sentence 'A cockerel is incapable of doing arithmetic', requires knowing the fact expressed by the verbal sentence; but getting to know that what it says is true, unlike learning that what the ontological sentence says is true, requires learning nothing in addition to the fact stated by the verbal sentence. It may be helpful to put the difference[17] between the verbal sentence and the nonverbal one by saying that the verbal sentence records what the nonverbal sentence, which mentions no word, presents. The likeness between the sentence with ontological content and the nonverbal sentence whose content is wholly verbal lies in the mode of discourse in which they are framed – what I have elsewhere called 'the ontological idiom'.

Making perspicuous the likenesses and differences between the three sentences is important because it brings out the fact that the verbal content of an utterance may be obscured by the form of speech in which the utterance is framed. It is possible for an ontologically formulated sentence whose content is verbal to create a delusive idea as to what the sentence is about. To illustrate briefly, the sentence 'A cockerel is a young male fowl' can be rewritten as an entailment-sentence: 'Being a cockerel *entails* being a young male fowl'. To many philosophers the entailment-sentence presents the appearance of expressing a proposition about essential properties of an object, properties the object cannot fail to have. These, as against nonessential properties which can be learned only by experience, are thought of as being discovered by analytical penetration into the concept under which the object falls. To see that this picture is delusive all we need do is translate the entailment-sentence back into 'A cockerel is a young male fowl', the content of which can be seen to be the fact stated by the verbal sentence. Seeing that the ontologically formulated sentence has verbal content evaporates a metaphysical theory.

It is important to notice that not only is a matter of established usage sometimes brought before us in the ontological idiom, but that re-edited terminology is frequently announced in it. Sometimes this is done merely for convenience, to avoid the more clumsy verbal form of locution. But sometimes it is done for the striking effect which presenting a redefinition in the ontological idiom is capable of producing. In a recent address to a Jewish audience the statement was made that assimilation is ethnic genocide. It is quite clear in this case that the word 'genocide' was being redefined, and that the application of the word was being *stretched* by an act of fiat so as to apply to whatever 'assimilation' is used to cover. The verbal sentence corresponding to the ontologically formulated sentence 'Assimilation is ethnic genocide' is, to put it roughly, 'Assimilation should be called "ethnic genocide" '. This certainly lacks the dramatic quality of the ontologically formulated utterance, which can produce the idea that an unsuspected property of assimilation is being disclosed. The way to dispel this idea is to show that the statement has only verbal content, viz., an arbitrary redefinition, which fact tends to be hidden by the ontological mode of speech.

According to the thesis to be developed in these pages a philosophical theory resembles in important respects the ontologically formulated statement about assimilation. Its content is verbal and it is capable nevertheless of creating the idea that it expresses a theory about things. The main difference between a philosophical utterance and the statement about assimilation is in the durability of the intellectual illusion it generates, which makes it more difficult to expose the underlying verbal content. In the case of an utterance like 'Motion is self-contradictory and exists only in appearance' or 'Ultimate reality is undifferentiated experience' the philosophic moment in which it is accepted as true or rejected as false undoubtedly has support from subterranean sources. This is to say that philosophical utterances give expression to unconscious clusters of ideas which serve to keep us at a distance from them. Nevertheless, the only way, to use Wittgenstein's word, of 'dissolving' the erroneous impression created by them is to expose the verbal content behind the ontological facade. In the case of the statement that motion is really impossible and exists only in appearance, what needs to be shown is that the word 'motion' is academically deprived of its use. Doing this consists in part of showing the difference between the sentence 'The images on the screen are not really in motion although they appear to be' and the philosophical sentence 'The arrow in flight is really stationary'.

The thesis developed here is that a philosophical theory is an illusion which is created by presenting in the ontological mode of speech a gerrymandered piece of terminology. It is a two-layer structure consisting, for one thing, of re-edited nomenclature, and for another thing, of the idea, which is generated by the way the nomenclature is introduced, that words are being used to express a theory instead of to herald a redefinition. Joined to these is a third and less accessible layer, a complex of unconscious fantasies. It will be maintained that unconscious ideas are given expression by a philosophical utterance; it is these which hold the philosopher in bondage to his theory and prevent him from seeing it for the meagre linguistic contrivance that it is. To allude again to Wittgenstein's metaphor of the fly in the fly-bottle, it is the unconscious fantasy that holds the philosopher captive and makes it so difficult to show him the way out. Nevertheless, the only way to lead him out is by laying bare the linguistic content of his philosophical utterances. Once this is done we can be free to conjecture regarding some of the unconscious material on which a philosophical structure rests.

One philosophical problem, which reaches all the way back to Parmenides, is linked in a special way with the notion of analysis as a method of investigating the world of things. It may be usefully considered first in the series of views to be examined. This problem revolves around the rationalistic claim that we cannot think of what does not exist. The two basic traditions in philosophy, empiricism and rationalism, appear to be antithetical, the one to all appearances accepting the senses as a source of knowledge of things, the other placing its reliance on reason alone, or at any rate the weight of its reliance on reason. There are good grounds for thinking that the difference between rationalism and empiricism is not, as it would seem, in their methods of investigation. There is reason to think that behind the scenes empiricism equally with rationalism rejects the employment of the senses in its investigations, and in the words of Plato 'examines existence through concepts'. Parmenides urged that we turn away from the senses and use 'the test of reason' in the investigation of reality; and if we can look behind the empiricists' talk, what we see is that under their semantic skin they are Parmenideans.

There are many and very different ways of saying the same thing in philosophy. There is even reason for thinking that the most sober of all philosophers, G.E. Moore, who explicitly protested that analysis was not the only thing he tried to do,[18] nevertheless was confined to the technique of analysis. One philosopher has supported Moore's protest, claiming that it was justified on the grounds that 'Moore also looked to philosophy to determine what kinds of things there were in the universe. . .'[19] But looking to *philosophy* to determine what exists is neither to look at things nor to experiment with them: the investigation can only be analytical and completely removed from things.

The claim that the philosophical empiricist is a disguised rationalist[20] cannot be gone into for the moment. The question of present concern is whether it is possible to think of what does not exist. A review of a recent book[21] contained the following statement: '. . . in Part II the problems chiefly discussed are those raised by the fact that what we think *about* may be *nonexistent*'. It is a curiosity, but one often met with, that what some philosophers declare to be a *fact* other philosophers declare to be impossible. Thus, Parmenides, and many other philosophers with him, took the position that it is impossible to think of what does not exist. One formulation of this position is the following: 'It is impossible to think what is not and it is impossible for what cannot be thought to be. The great question, *Is it or is it not?*, is therefore equivalent to the question, *Can it be thought or not?*'[22]

One consideration in support of the proposition that is impossible to think of what does not exist is that thought must have an object. To think is not to think of nothing; it is to think of something, which therefore must exist, as not to exist is to be nothing. Wittgenstein put the matter in the following way: 'How can we think what is not the case? If I think that King's College is on fire when it is not on fire, the fact of its being on fire does not exist. Then how can I think of it? How can we hang a thief who does not exist?' He goes on to say, 'Our answer could be put in this form: "I can't hang him when he doesn't exist; but I can look for him when he doesn't exist" '.[23] A Parmenidean rejoinder would no doubt be that we cannot *really* look for a thief who does not exist. The object of our search and of our thought equally with the object of our act of hanging him cannot be a thief who does not exist.

Some philosophers who explicitly reject the rationalistic claim of Parmenides seem nevertheless to reinstate it in a more subtle form, under the cover of a puzzling distinction. Thus, Bertrand Russell, who rejected the existential theory of judgment, viz., that 'every proposition is concerned with something that exists',[24] has written: 'Whatever can be thought of has being, and its being is a precondition, not a result, of its being thought of. As regards the existence of an object of thought, however, nothing can be inferred from the fact of its being thought of, since it certainly does not exist in the thought which thinks of it.'[25] He observed that 'Misled by neglect of being, people have supposed that what does not exist is nothing'.[26] The term 'being' has two uses which are relevant here. In one use it has the meaning of a kind of thing or entity, and in this use it is correct to speak of the Greek gods as beings. In another use it has the meaning of existence, and in this use it is, in the ordinary way of speaking, correct to say that the Greek gods have no being, i.e., do not exist. It does not seem too much to suppose, for the present at least, that confounding these two senses of 'being' is responsible for the philosophical distinction between the terms 'has being' and 'has existence', which serves to reinstate in a subtly disguised form a claim that is explicitly rejected. We may say that some things which exist are beings and that other things which exist are not beings, e.g., shadows and sneezes, but that everything that exists has being. Whatever has being (regardless whether it is a being or a kind of thing) is part of what there is and exists. Hence the claim that whatever can be thought of has being implies the Parmenidean proposition that whatever can be thought of exists. Russell remarked that 'Everyone except a philosopher can see the difference between a post and my idea of a post',[27] and there is reason for thinking that only a philosopher could see a difference between saying that a thing has being and saying that it exists.

F.H. Bradley, the Oxford Parmenides of the twentieth century, appears to have held that we cannot think of the nonexistent. The following two remarks suggest this as clearly as anything else he says in his *Appearance and Reality*: '. .to suppose that mere thought without facts could either be real, or could reach to truth, is evidently absurd';[28] and 'A mere thought would mean an ideal content held apart from existence. But (as we have learnt) to hold a thought is always somehow, even against our will, to refer it to the Real'.[29] Parenthetically, it is worth noting that underlying the so-called consistency theory of truth is the notion that conceivability and existence coincide, i.e., that the existence and nature of things can be read off from our self-consistent thoughts, and the nonexistent from our self-contradictory thoughts. G.E. Moore, who rejected the Parmenidean thesis, nevertheless constructed an interesting argument for it. He wrote:

How (I imagine he [Bradley] would ask) can a thing 'appear' or even 'be thought of' unless it is there to appear and to be thought of? To say that it appears or is thought of, and yet that there is no such thing, is plainly self-contradictory. A thing cannot have a property, unless it is there to have it, and since unicorns and temporal facts *do* have the property of being thought of, there certainly must be such things. When I think of a unicorn, what I am thinking of is certainly not nothing; if it were nothing, then, when I think of a griffin, I should also be thinking of nothing, and there would be no difference between thinking of a griffin and thinking of a unicorn. But there certainly is a difference; and what can the difference be except that in the one case what I am thinking of is a unicorn, and in the other a griffin? And if the unicorn is what I am thinking of, then there certainly must be a unicorn, in spite of the fact that unicorns are unreal.[30]

Moore thought it obvious that this argument contained a fallacy and he suggested that people might even think it 'too gross' for Bradley to have been guilty of it — although he was quite sure that Bradley was guilty of it. An outsider to philosophy, to whom it is represented as a highly rational discipline, one that uses exact analytical tools, might be shocked to witness one known and important philosopher declare the view of another known and important philosopher to be a gross error. He would be even more shocked, and astonished, to learn that the mistake has been in existence, and never without adherents, for a truly remarkable number of centuries. In philosophy 'grossly mistaken' claims seem to have an infinite viability: what is a plain falsehood to some people remains a clear and indestructible truth to others, for generation after generation. We are, to be sure, familiar with this kind of situation in religion, where it is not considered to be an enigma. But to discover that it exists everywhere in a discipline which parades as a *demonstrative science* is to be faced with an enigma that challenges our intelligence. To leave this for the moment, however, it is interesting, and important, to notice that although the Parmenidean-Bradleian 'mistake' is crass, Moore confesses to not being sure what the mistake is in the claim that our being able to think of a unicorn is sufficient to prove that there is one.

As it seems to him, the mistake, or at any rate the main mistake, consists of supposing that the statement 'I am thinking of a unicorn' is of the same form as the statement 'I am hunting a lion', whereas in fact they are different in a crucial way. The second statement is equivalent to 'There are lions (at least one) and I am hunting it'.

$$(\exists x) (x \text{ is a lion. I am hunting } x),$$

while 'it is obvious enough to common sense' that the first statement has no like equivalent, although 'their grammatical expression shows no trace of the difference'.[31] A philosopher always has a rejoinder (which is perhaps what made Moore not quite certain that the 'mistake' was a mistake), and a Parmenidean would respond that the philosopher of common sense was begging the question in claiming that the statement 'I am thinking of a unicorn' does not entail the statement 'There are unicorns'. It would be urged that instead of meeting the argument, which supports the thesis that the existence of the thought of a thing implies the existence of a thing corresponding to the thought, Moore actually does no more than assume the counter-thesis. It is easy to expand on this defense against Moore and say, e.g., that if the analysis of 'I am hunting a lion' into a conjunction of statements *is* correct, then that is also the correct analysis of 'I am thinking of a unicorn',

$$(\exists x) (x \text{ is a unicorn. I am thinking of } x).$$

Wittgenstein's remark that although we cannot hang a thief who does not exist, we can look for one who does not exist, which he took to be an *answer* to the Parmenidean claim, would also be rejected as a *petitio*. The rejoinder would be that just as I cannot hang a thief who does not exist, so I cannot look for him nor yet think of him. If I could hang him, he must exist; if I can look for him, he must exist; and also, if I can think of him, he must exist. The observation that is relevant at this point is that if the words 'It is possible to think of what does not exist' give the correct answer to the question 'Is it possible to think of what does not exist?', then a Parmenidean knows this as well as did Moore and Wittgenstein. The important question to investigate is the nature of the Parmenidean thesis. Do the words 'Whatever can be thought of exists', or the words 'It is impossible to think of what does not exist', express an experiential proposition or do they express an *a priori* proposition or are they intelligible but express neither an *a priori* nor an experiential proposition?

There is no question about the kind of claim the words *seem* to express. They give every appearance of making a statement about what we can think and about how things are related to our thoughts. In fact so vivid and lively is this appearance that to deny, or even doubt, that they do make such a statement would seem to indicate an impairment in one's sense of reality. Nor is it to be doubted that, at the conscious level of their mind, it is the idea that the words make a factual claim about thoughts and things which wins and holds the attention of philosophers, both those who reject the idea as false as well as those who embrace it as an ontological truth. This idea is as self-evident and compelling as once was the idea that the earth is flat. The strange thing is that the known points of detail which upset the idea about the earth and enabled Aristarchus to calculate the earth's circumference to a remarkable degree of accuracy were thrown into shadow and forgotten. Undoubtedly it was the alarming astronomical picture associated with the new idea which required that it be lost to memory. In philosophy also small known points of detail which, when investigated, lead to disconcerting revelations have continuously remained in the shadow of inattention. One of these is the curious intractability of philosophical disagreements, in the present instance the truly strange intractability of the disagreement over whether it is possible to think of what does not exist. Both Descartes and Kant expressed their dissatisfaction with the total anarchy of opinions in philosophy, but neither turned out to be an Aristarchus.

If we reflect with detachment on the words 'Whatever we think of exists', we will realize that, however a philosopher is using them, he is not using them to state a claim that is founded on experience or to which experience is in any conceivable way relevant. If, like the words 'Whatever we wish comes to be', their use was to describe an alleged state of affairs, the disagreement over whether the state of affairs actually obtains would have been resolved long ago, if indeed it would ever have come into existence. A person who holds that it is impossible to think of what does not exist has not come to his notion as a result of *trying* in a number of instances to think of something that does not exist and failing in each instance. And a philosopher who declares that whatever we think of exists has not been taught by experience that for every thought of a thing there exists a thing. His statement is not an inductive generalization from observed concomitances of thoughts and things: the thought of an elephant with an actual elephant, the thought of a dolphin with an actual dolphin, the thought of a buttercup with an actual buttercup, etc.

A generalization which issues from a series of concomitances that experience discovers, such that the dissociation of any of the observed concomitances is possible in principle, is one which, no matter how secure experience shows it to be, is nevertheless a generalization to which an exception is conceivable. If we look on the claim that every thought has a corresponding real object as an inductive generalization, we are faced with a bewildering puzzle: what some philosophers cite as confuting instances, as cases which upset the generalization, other philosophers maintain are *not* confuting instances. A philosopher who cites as an exception his thought of a purely imaginary creature, such as a unicorn or a dragon, finds that his example is rejected as not really going against the general claim. The difference of opinion amongst philosophers over the putative counterexample appears itself to be factual. If, however, we examine it with care we can see that the generalization is not empirical and neither is the disagreement over the example.

To suppose otherwise, that is, to suppose the difference of opinion about whether an actual unicorn corresponds to the thought of a unicorn is factual, leaves us puzzled to know what could possibly resolve the dispute. For the experience of the philosopher who insists that there is a unicorn is no different from that of the philosopher who denies that there is. The one does not have a delusive perception of a creature which does not exist, nor does the other fail to perceive a creature that does exist, although the factual interpretation implies either that the Parmenidean philosopher suffers from something comparable to a hallucination or that the commonsense philosopher suffers from a kind of selective blindness. We can hardly be criticized for being reluctant to accept this implication about the minds of philosophers on either side of the issue. But if we reject the implied consequence of the natural interpretation of the Parmenidean view, we also have to reject the construction it is natural to place on the dispute.

We approach the correct, paradox-free understanding of the words 'You cannot think of a unicorn that does not exist' and of the general statement that it is impossible to think of what does not exist, when we cease confining our attention to a single disputed example. We then realize that *every* instance cited as being a case of someone thinking of what does not exist would be rejected as being such an instance and is instead declared to be a case of thinking of what does exist. If for the sake of the argument we allow that we are mistaken in supposing that unicorns do not exist and that dinosaurs no longer exist, and go on to ask our philosophical adversary for a *hypothetical* instance of someone thinking of what does not exist, we shall find him unable to provide us with one. He cannot say what it would be like to think of what does not exist. With regard to an accepted generalization like Bernoulli's law that an increase in the velocity of a fluid is regularly accompanied by a decrease in the pressure it exerts, it is in principle possible to *describe* an instance which, if it existed, would upset the generalization. An instance of the velocity of a fluid being increased without the occurrence of a corresponding decrease in its pressure is conceivable even though this never in fact happens. This shows what might be called a logical difference, a difference in kind, between the philosophical generalization and the scientific one, and it helps us see in an improved light the nature of the disputed instance. It helps us see that a philosopher who insists that unicorns exist is not under the domination of an overstrong imagination.

The fact that the philosopher cannot give us an idea of what he would accept as upsetting his general claim shows that the philosophical claim that whatever we can conceive has real existence, unlike Bernoulli's law, is not open to falsification by any theoretical experience. It is not an experiential proposition, however much the language expressing it suggests. And that it is not a proposition which could be either confuted or supported by observation or by any experiment fits in with the kind of evidence adduced in its support, an argument. The words 'I am thinking of a nonexistent unicorn' or the words 'I am thinking of the highwayman Dick Turpin who no longer exists' are rejected not a presenting false empirical claims but as stating what is logically inconceivable.

A classical scholar remarked that an ancient view, according to which there was neither motion, nor change, nor variety of phenomena in the world, flouted all experience and was supported by an argument in which evidence played no role. In his opinion the view is 'a way of thinking about things which is perpetually refuted by actual contact with things'.[32] As is known, Moore has said a similar thing, to which philosophers have paid no discernible heed.[33] We should indeed think a person to be suffering from an unusual state of mind who held beliefs about things which are 'perpetually refuted' by his actual experience of them or who held views about things which he knew to be false. And we should have to think philosophers a strange breed who were able actually to believe that one cannot think of what does not exist, of a leprechaun at his writing desk, of a winged hippopotamus, or of a friend who is no more. However, the strangeness surrounding the philosophical view, which is held solely on the basis of an argument regardless of sense-evidence, is largely dissipated once we realize that it is not empirical and thus cannot go against any experience, actual or possible. As the philosopher uses the words 'No one can think of what is not', whatever their literal import may turn out to be, they do not express an empirical proposition, one which flies in the face of everyday experience. A philosopher who states that it is impossible to think of what is not the case does not recognize that his statement is perpetually refuted by 'actual contact with things', not because he fails to see a glaring inconsistency with his experience but because it is not the kind of statement that experience could conceivably refute. It cannot be upset (or confirmed) by experience because it is not about a possible experience, about what we can and cannot think. The philosopher does not recognize that his statement is inconsistent with what he knows to be true, namely, that he can and frequently does think of what is not or of what no longer is, because it is not inconsistent with what he knows.

The problem which faces us is how to understand the philosopher's words. Once we give up the notion that they are used to make an empirical claim, to the testing of which experience is relevant, it seems entirely natural to adopt the notion that they make an *a priori* claim about what we can and cannot think. Seen in this way, they appear to state a proposition which is like that expressed by the words 'It is impossible to think of an oblong with no more than three sides' or by the words 'It is impossible to ride a horse that does not exist'. We are accustomed in philosophy to the distinction that is made between the accidental properties of a thing, properties which it could be conceived of as not having, and its essential properties, without which it cannot be conceived. We are also familiar with the philosophical distinction between necessary existence and contingent existence.[34] Important philosophers have maintained, for example that a perfect being possesses its existence by *a priori* necessity, as against mice and planets which exist contingently. And some important philosophers have maintained that being such that its existence is independent of its being perceived is an essential property of a material thing. Moore's way of putting this was that we should not 'call anything a material object' whose existence was dependent on its being perceived.[35]

The point of remarking on these well-known distinctions is to call atten-.tion to the notion which all philosophers have, even those who make a point of rejecting it, that it is possible for a proposition to be *a priori* and also about things. Thus, the words 'A perfect Being exists by the necessity of its own nature' are understood by philosophers to make an *a priori* claim about the existence of a special object. Similarly, the words 'A material thing exists independently of its being perceived' are taken to express an *a priori* proposition about the nature of material things. And when it is realized that the sentence 'it is impossible to think of what does not exist' (and the equivalent sentence 'Whatever is thought of exists') does not state an empirical proposition, the idea that it expresses a necessary proposition about how thought and reality stand to each other presents itself with irresistible force. There is no gainsaying the powerful impression the sentence creates of being about thoughts and objects, an impression so vivid that, to use Kant's words, 'even the wisest of men cannot free themselves from it'. This together with the fact that an argument is brought in to support the claim it makes gives rise to the notion that the claim is both nonempirical, or *a priori*, and about things. The impossibility referred to by the sentence 'It is impossible to think of what does not exist' is a logical impossibility; and the necessity referred to by the corresponding sentence, 'It is necessarily the case that whatever one thinks of exists', is a logical necessity.

Many, if not all philosophers,[36] are under the domination of the impression that logical necessity and physical necessity, and also logical impossibility and physical impossibility, are generically the same, and that they differ from each other only in degree of inflexibility. Thus, Wittgenstein speaks of 'the hardness of the logical *must*'.[37] The usual view is that a logical impossibility implies a corresponding physical impossibility and a logical necessity the corresponding physical necessity, but not conversely. What is physically impossible, i.e., an impossibility in nature, cannot in fact be upset, but can be conceived of as being upset; whereas what is logically impossible can be upset neither in fact nor in conception. And while a physical necessity, i.e., an immutability laid down by a law of nature, cannot be overruled in fact but can in conception, a logical necessity cannot be overruled either in fact or in conception. One of Russell's remarks illustrates this view graphically. He wrote: 'Men, for example, though they form a finite class, are, practically and empirically, just as impossible to enumerate as if their number were infinite.'[38] It is, of course, logically impossible to finish counting a non-ending number of elements, as this would imply counting the last member of a series which has no last member. This fact in conjunction with Russell's remark would imply that as in the case of an astronomically huge set of elements which it is 'practically and empirically impossible to enumerate, the elements of an infinite aggregate cannot be enumerated either empirically or in conception. The idea is that running through the elements of such an aggregate is not only logically impossible but is *also* physically impossible. This is the picture we naturally form of the relationship between the concepts of logical and physical impossibility. Our picture changes radically, however, when we scrutinize these concepts with care.

Consider the following two sentences:

'It is impossible to grow a tulip which is not a flower',
'It is impossible to grow a tulip from an acorn'.

Russell would undoubtedly say that it is just as impossible 'empirically' to grow a tulip from an acorn as it is to grow a tulip that is not a flower. It is clear, however, that the impossibility described by 'grows a tulip which is not a flower' is an inconceivability, as against the impossibility described by 'grows a tulip from an acorn', which is conceivable. Following Spinoza, we might say that it would take God to make a tulip grow from an acorn, but that not even God could grow a tulip which is not a flower. Furthermore, not even God could conceive of something being a tulip and not a flower. This means that God would attach no descriptive sense to the phrase 'grows a tulip that is not a flower', which is a theological way of saying that the phrase has no descriptive sense. The implication is that unlike the phrase 'grows a tulip from an acorn', the function in the language of 'grows a tulip that is not a flower' is not to describe anything. The expression which refers to what is declared to be impossible in the sentence 'It is impossible to grow a tulip from an acorn' *describes* what is declared to be impossible, while the corresponding expression in the sentence declaring a logical impossibility does not describe a circumstance that is declared impossible. This holds for the two sentences, 'It is impossible to enumerate the men in the world' and 'It is impossible to enumerate an infinite number of elements'. The referring phrase of the first sentence describes a process, while the referring phrase of the second does not. This difference between logical and physical impossibility shows that they are different in kind. It also shows that a sentence which makes a statement as to what is logically impossible is not about things. The Parmenidean sentence 'It is impossible to think of what does not exist', construed as stating a logical impossibility, says nothing about what cannot be thought. Taking the word 'impossible' to mean 'logically impossible', the phrase 'thinks of what does not exist' has no use to describe what we cannot think.

Another way of seeing that a sentence which expresses a logically necessary proposition says nothing about things is provided by considering the affirmative equivalent of 'It is impossible to grow a tulip that is not a flower', namely, 'To grow a tulip is necessarily to grow a flower', or to simplify the example, 'A tulip necessarily is a flower'. It will be clear that the term 'flower' does not function in the sentence to distinguish, either in fact or in conception, among tulips. The adjective 'yellow' in 'A buttercup is always yellow' functions in the sentence to distinguish among conceivable buttercups. It sets off buttercups that are yellow from other imaginable buttercups and thus has a characterizing use. The term 'flower' does not set off tulips that are flowers from other possible tulips, and therefore, does not have a characterising use in the sentence 'A tulip necessarily is a flower'. Whatever the information the sentence may be used to convey, it has no use to convey information about tulips. The point may perhaps be brought out by rewriting the two sentences, in somewhat unnatural English, in the following way: 'A tulip is a tulip that is a flower', 'A buttercup is a buttercup that is yellow'. The expression 'a tulip that is a flower', unlike the expression 'a buttercup that is yellow', attributes nothing to what is referred to by its grammatical subject; the meaning of 'a tulip that is a flower' is not something in addition to the meaning of 'tulip'. Thus, with respect to what it asserts about *a property* of tulips, there is no difference between 'A tulip is a tulip that is a flower' and the factually empty tautology, 'A tulip is a tulip'. The first sentence is as empty of information about tulips as is the second. Neither sentence states anything about the nature of tulips, and we might say, therefore, that neither is *about* tulips. To make a general statement regarding sentences which express *a priori* propositions, they are all empty of information about the world. They are ontologically mute.

This general maxim is of first importance for a correct understanding of the way philosophy works. It can readily be seen to apply to the words 'Whatever thing we think of must exist', providing they express, or are advanced as expressing, an *a priori* truth. Consider the following equivalent even though awkward way of writing the sentence, 'Whatever thing we think of exists': 'It is always the case that the thought of a thing is the thought of an existing thing'. Since, by hypothesis, the sentence expresses a logically necessary proposition, the phrase 'the thought of an existing thing' does not, as does the phrase 'the thought of a buttercup', function in such a way as to set off some thoughts from other thoughts, e.g., thoughts of buttercups from thoughts of things which are not buttercups. It does not serve to set off thoughts of existing things from thoughts of nonexisting things. This implies that the *theoretical* range of application of the two expressions, 'thought of an existing thing' and 'thought of a thing', coincide precisely. This in turn implies that the first says no more about thoughts than does the second. With respect to thoughts, they say the same. Hence, viewed as expressing a logically necessary truth, the sentence 'All thoughts of things are thoughts of existing things' is as barren of information about thoughts as is the tautology 'All thoughts of things are thoughts of things'. Wittgenstein declared that a tautology 'says nothing'.[39] Not to go into the question as to whether a tautology says nothing at all, we can say that a tautology says nothing about *things*. Whatever it is that we know in knowing a proposition to be a tautology, or more generally an *a priori* truth, we know nothing about the world.

It is relatively simple now to see why a logically necessary truth cannot imply a corresponding factual truth and why a logical impossibility cannot imply a corresponding physical impossibility. The general reason is that from a proposition which carries no information whatever about things no proposition which supplies such information is deducible. A proposition which has factual content cannot be entailed by a proposition which has none. The proposition expressed by 'It is physically impossible for a kangaroo to jump a mile' states what a kangaroo cannot as a matter of fact do. The phrase 'kangaroo which jumps a mile' describes what we can imagine happening in nature, although it never in fact happens. Hence the proposition cannot logically be inferred from any necessary proposition, any more than it can be inferred from the proposition that a kangaroo either jumps a mile or does not jump a mile. It is the same with the proposition that bread must bake at a temperature of 400 °F. It is tempting to adopt the view that what is true necessarily is also true as a mere matter of fact and that what is logically impossible is also physically impossible. But taken for what it appears to be on the surface, as making a logical claim about entailments between kinds of propositions, the view is mistaken.

Perhaps the best way of making this clear is to articulate the logical point that is involved. The fact that a given proposition has its truth-value by logical necessity prevents it from being about things or from having factual content; and the fact that a given proposition is about things implies that it does not have its truth-value by logical necessity. The general point is that an *a priori* proposition cannot entail one that is contingent. A proposition whose actual truth-value is its only possible truth-value cannot entail a proposition which is in principle capable of either of two truth-values. To put the matter with the help of C.I. Lewis' symbol for logical possibility, '$\Diamond$', where we have p necessary and q possibly false, i.e.,

$$\sim\Diamond\sim p, \quad \Diamond\sim q,$$

we cannot have

$$p \rightarrow q.$$

For the theoretical possibility of q having a truth-value other than the one imposed on it by its being entailed by p is ruled out by p's being necessary. The consequences of an *a priori* proposition are themselves *a priori* and thus empty of ontological content. Interpreted as denoting a logically necessary proposition, the sentence 'Whatever is thought of exists' will fail to have an associated sentence expressing a matter of fact claim about what invariably happens in nature.

Two connected questions present themselves at this point. One is the general question as to what necessary propositions are about, or better, what the subject-matter of a sentence expressing a necessary proposition is. The subject-matter of the sentence 'Percherons are draft horses' is Percherons, horses of a certain breed; but philosophers seem to wander in a fog, either Platonic or nominalistic, when they seek to identify the subject-matter of the sentence 'A Percheron is an animal'. The other question is whether the philosophical sentence, 'It is impossible to think of what does not exist' (and its affirmative equivalent), does in fact express a logically necessary proposition and whether it is put forward by Parmenidean philosophers as doing so in ordinary language. There is a tendency to think that it does make a nonempirical claim, and despite disavowals, philosophers cannot resist thinking that the claim is both nonempirical and also about things. Kant had the notion that a proposition could have 'inner necessity' and also be a truth about phenomena. This notion has been taken by many philosophers as the rock upon which the metaphysical science of the world would be built. As has been seen, however, if a proposition is *a priori* it cannot carry information about phenomena, regardless of whether or not its predicate is a component of its subject. The plain implication is that if one wishes to read a book of the world, no book in the vast library of philosophy will be of the slightest use. Hume stated this point colorfully when he said that if a book contains no abstract reasoning concerning number nor any experimental evidence concerning matter of fact, 'it can contain nothing but sophistry and illusion'.[40]

It is important for the light thrown on philosophical theories to get a clear idea of what necessary propositions are about and of the kind of information they carry. Necessary propositions, one would gather from the abundance of philosophical theories about their nature, are a baffling and elusive species. To mention several of the theories: one has it that a necessary proposition is an inductive generalization to which a high degree of probability attaches; according to another theory, it states an unalterable relationship between abstract entities; a third maintains that it is about the use of terminology in a language. It is bewildering, to say the least, to witness three learned people reflecting on a simple proposition such as that a Percheron is an animal, or that $2 + 2 = 4$, and emerging with intractably rival conclusions: one, that the arithmetical proposition is about the terms '2 + 2' and '4', another, that it is about abstract entities denoted by these terms, and another, that it is an inductive proposition based on the examination of instances. There are still other theories. It would not, however, be to the point here to venture into the maze of philosophical theories concerning the *a priori*, however inviting that may be. The question as to what necessary propositions are about and what kind of information we obtain from them is dealt with best and with the least amount of mystification by looking at them, as before, through the spectacles of the *sentences* used to express them. And the explanation will be least roundabout if we consider sample sentences which express propositions stating something to be impossible.

The sentence 'It is impossible for there to be a Percheron which is not an animal', like the sentences 'It is impossible for there to be 4 minus 5 sheep in the meadow' and 'It is impossible to grow a tulip which is not a flower', makes no statement about what is not the case. The fact that it expresses a proposition which is not open to falsification implies that the phrase which refers to what is declared by the sentence to be impossible has no use to describe an imaginable thing or occurrence or state of affairs. The phrase 'Percheron which is not an animal', equally with 'keeps 4 minus 5 sheep in the meadow', 'comes to the last member of an unending series', and 'grows a tulip which is not a flower', is *descriptively senseless*. Part of *understanding* the sentences 'It is impossible for there to be a Percheron which is not an animal' and 'It is impossible to keep 4 minus 5 sheep in a meadow' consists in knowing that the phrases referring to what is declared to be impossible have no descriptive content. Hence, knowing that these sentences express *a priori* propositions is equivalent to knowing that certain combinations of words in a language do not function descriptively in that language. It is this fact which leads some philosophers to the view that necessary propositions are really verbal. The objections to this view are well known and would seem to be conclusive, although knowing these objections does not deter philosophers from holding it. We may be sure that what makes the conventionalist secure in his view about the nature of *a priori* necessity in the face of the objections is that his view is *philosophical*. We have only to remind ourselves of Moore's paradox to realize this.

The subject-matter of the sentence 'It is impossible for there to be a Percheron which is not an animal' is no more verbal than it is factual. The fact that in knowing that the sentence expresses an *a priori* truth what we know is an empirical fact about a combination of words, and furthermore that this is all there is to know, creates the temptation to identify this fact of usage as the subject-matter of the sentence. But the sentence makes no declaration about the phrase occurring in it. The fact that the phrase has no descriptive content prevents the sentence from being about things, and the fact that the sentence expresses a logically necessary proposition prevents it from making an empirical declaration bout usage. The consequence, however strange it may at first seem, is that the sentence, even though perfectly intelligible, has no subject-matter: it says nothing about either things or words and thus, to use again Wittgenstein's expression, says nothing.

The idea that an indicative sentence might have a literal meaning but nevertheless have no subject-matter, and make no statement about anything, apparently produces discomfort in the minds of philosophers. Like an oyster which tries to remove an irritant by manufacturing a pearl, a philosopher tries to make up for what language denies him by *manufacturing* a subject of discourse for it. The result is a thicket of theories. Philosophers of the *a priori* rid themselves of a linguistic discomfort but in doing so produce a typical philosophical symptom, an irreducible number of views. Anyone who prefers fact to semantically induced fantasy will experience little trouble reconciling himself to the idea that a sentence whose meaning is an *a priori* proposition has no subject about which it makes a statement. He will have little trouble seeing that, although it says nothing about words, in knowing what is says we know only facts of usage. This fact about sentences for *a priori* propositions is, perhaps, shown most simply and clearly by writing out the following equivalences:

(1)   the fact that the sentence 'It is impossible to grow a tulip which is not a flower' expresses an *a priori* proposition

is equivalent to

the fact that the sentence ' "grows a tulip which is not a flower" is a phrase that has no descriptive function' expresses a true verbal proposition.

(2)   the fact that the sentence 'A tulip is a flower' expresses a logically necessary proposition

is equivalent to

the fact that the sentence ' "flower" applies by reason of usage in the language to whatever "tulip" applies to' expresses a true verbal proposition.

To hold that (1) and (2) are equivalent is not to imply that the proposition expressed by 'The phrase, "grows a tulip which is not a flower" has no use' is the same as the proposition expressed by 'It is impossible to grow tulip which is not a flower'. Neither is it to imply that the proposition expressed by 'Usage dictates the application of the term "flower" to whatever "tulip" applies to' is the same as the proposition expressed by 'A tulip is necessarily a flower', or by 'Being a tulip *entails* being a flower'. The fact about usage expressed by the first sentence of each of the two pairs of sentences is what we know in understanding the second sentences, and we know nothing in addition to these facts in knowing the propositions expressed by the second sentences. The facts which we know are nevertheless not the subject-matter of the sentences and cannot be identified with the propositions they express. Mathematics, which may be truly described as the systematic science of the *a priori*, has been a source of wonder and mystification to many people. G.H. Hardy represented the mathematician as gazing into an intricate system of objects less gross than those encountered in sense experience, and recording what he sees. The view we have arrived at here about the nature of the *a priori* de-Platonizes mathematics, without reinstating Hilbert's form of conventionalism. To make an observation on *part* of what a mathematician's work consists in, it is the explication of rules of usage presented in a form of speech in which no terms are mentioned, i.e., in the ontological idiom.

We are now in a position to see clearly that the sentence 'It is impossible to think of what does not exist' does not, as the words entering into it are ordinarily used, express an *a priori* proposition. The supposition that it does implies that the phrase 'thinks of what does not exist' is not a descriptive combination of words, that it has no use to convey information. And the supposition that the equivalent sentence 'Whatever is thought of exists' expresses a logically necessary proposition implies that usage dictates the application of 'exists' to whatever is named as the object of thought: usage dictates the application of 'exists' to what is meant, for example, by the nouns in the expressions 'thinks of a nightingale' and 'thinks of a Gorgon'. It will be plain that anyone who has the idea that these sentences put forward logical claims will have the idea that 'thinks of what does not exist' has no application to any theoretical occurrence and that 'exists' applies to whatever is referred to by terms which are intelligible values of x in 'thinks of x'. To anyone who is able to resist playing the game of philosophy, it will also be plain what the facts with regard to usage are. It is a fact that the phrase 'thinks of nonexisting things' does have a correct use in the language, and it is a fact that there is no rule dictating the application of 'exists' to whatever is referred to by suitable values of 'thinks of x'. To touch briefly on the term 'nonexisting thing', it and the term 'existing thing' constitute a pair of antithetical terms which semantically lean on each other. If one is deprived of its use, the other loses its use. If 'nonexisting thing' did not serve to distinguish amongst describable things, 'existing thing' would not serve to set off some describable things from others. Hence it could not function descriptively in the language. From the supposition that 'A thought of a thing is a thought of an existing thing' expresses a logically necessary proposition it follows that the sentence is tautological, that is, empty of content, with respect to thoughts of things; and being able to show that it follows hinges on the fact that 'existing thing' and 'nonexisting thing' are antithetical terms.

As is known, many important philosophers have, if we are to take them at their word, fancied that they had demonstrated contradictions in the meanings of such everyday words as 'motion', 'cause', 'space', and 'time', and some recent language-oriented thinkers have gone straightaway to the position that these philosophers had false beliefs about the use such words have in ordinary language.[41] A paradoxical claim like 'Motion is self-contradictory, and does not exist' is capable of presenting a linguistic face to some philosophers, just as it is capable of presenting an ontological face to others. But if we notice the talk and behavior with which the claim is surrounded, the impression that it implies a mistaken idea about actual usage evaporates. It is not necessary to look hard in order to see that there is nothing in the behavior of a metaphysician who says motion is self-contradictory to suggest that he actually believes people who use motion-indicating words to be suffering from a linguistic delusion, the delusion that they are conversing intelligibly while making no literal sense. Nor is it difficult to see that the metaphysician himself uses motion terminology in the ordinary conduct of life in the way everyone else does. Neither in his talk nor in his behavior, nor even in his behavior while pronouncing and arguing for the philosophical view that motion is unreal, does he betray having unusual thoughts about usual terminology. His view seems to be completely dissociated from his everyday use of language. This is why Bradley was able to say, 'Time, like space, has most evidently proved not to be real, but to be a contradictory appearance. I will, in the next chapter, reinforce and repeat this conclusion by some remarks on change.'[42] The reason why the inconsistency in these words went unnoticed, or if noticed soon fell under amnesia, is that there is no inconsistency. This means that a paradoxical philosophical theory (as well as a superficially non-paradoxical theory) is not a disguised statement about the correct use of an expression.

It is now generally recognized, except by the most benighted, that after being exposed as mistaken a philosophical view returns as a claimed truth, sometimes in its original form, sometimes dressed up in new terminology. This occurs over and over again. A philosophical 'mistake' is like the giant

Antaeus who sprang back with renewed vigor every time he was sent crashing to earth by Hercules. Antaeus finally met his end, but a philosophical mistake is never laid to rest. A mistake that occurs in scientific disciplines does not behave in this way; and the realistic conclusion to draw, which fits in with the behavior and ordinary talk of philosophers, is that a philosophical 'mistake' is not a mistake. An American general, after being removed from his command, said that old soldiers do not die, they just fade away. A philosophical mistake may for a time fade away, but it never goes out of existence and never is in want of adherents. The *déjà vu* feelings we so often experience in philosophy invariably turn out to be justified when we succeed in translating the new claim back into an old and familiar one.

A philosopher presents his theory with the air of someone who brings light and truth to his special subject, but the words he uses make no reference whatever either to things in the world or to the actual use of terminology in a language. What the nature of his theory is, or, better still, what he is doing with terminology, requires an explanation which will fit in with the fact that philosophy never frees itself from its mistakes, and therefore can never look forward to the resolution of disputes which everywhere infect it. The only explanation which does this is that his theory is a disguised introduction of changed terminology (or sometimes a disguised counter to it). A philosopher does not use familiar language either in the way the natural scientist uses it or in the way the mathematician uses it. His work consists of doing nothing more substantial than altering language; but he presents his verbal creations in the fact-stating form of speech and thereby conceals from all of us, including himself, what he does with terminology. Wittgenstein perceived this and was led to remark: 'Die Menschen sind im Netz der Sprache verstrickt und wissen es nicht'.[43] A debate over a gerrymandered piece of standard terminology can go on and on, without prospect of future resolution, because no sort of *fact* is in question. It will be clear that a 'mistake' which is constituted by redistricted nomenclature is in no real danger of being refuted by fact and is capable of an indefinite number of reincarnations. Wittgenstein said that a philosophical problem has no solution, but only a dissolution. This is a way of saying that a philosophical question has only to be understood aright, be linguistically unmasked, in order to cease being a *problem*. The idea that a philosophical theory is an ontologically presented language-innovation (or an ontologically expressed opposition to it) helps us understand why disputes over whether it is true or false are intractable. They are disputes in which exotic linguistic preference, not truth, is at issue. The fact that the disputes are endless also indicates that no practical consequences, which normally justify the adoption of a new notation, attach to a language change introduced in philosophy. It is noticing this which made Wittgenstein remark that philosophical language is 'like an engine idling, not when it is doing work'.[44]

An idle semantic departure from actual usage when presented in the form of speech in which statements about things are made is capable of generating a dispute of indefinite duration. No fact, old or new, can bring it to a final conclusion. And one thing, but not the only one, which makes it important to philosophers and gives it continuing life, is the appearance it has of being a dispute about things or occurrences in the world. Even so-called linguistic philosophers remain the dupes of this appearance. Their idea is that in some way the investigation of linguistic usage is relevant to a philosophical dispute about things, and in the end will settle it. The notion behind this idea is that a false philosophical theory is a disguised misdescription of usage, but that a true philosophical theory is not just about words but is about the world. The implication which is imbedded in this notion is that there can be no false philosophical theory about *things*, and thus that a true philosophical theory is a necessary truth about things. This can be recognized as a reincarnation of the idea that truths about things can be learned without going to the things themselves.

It is easy to understand now what it is about such sentences as 'Motion does not really occur' and 'To be a thing, e.g., a glove, is to be perceived', which creates the impression that they are about reality. Quite generally, a gerrymandered piece of terminology when presented in the fact-stating form of speech tends to create the impression that a theory is being stated, and this impression is strengthened to the point of becoming irremovable when the expression of a hidden idea is involved.[45] What Moore called a philosophic moment is compounded of two symbiotic ingredients: a piece of trumped-up semantics in union with a temporary suspension of the reality principle[46] in favor of an unconscious wish. A philosopher who explicitly or by implication holds that a truth about things can be obtained without investigating them, at a superficial level plays a game with language but at a deeper level uses the game to gratify the still active desire for omniscience. In the case of the Spinozistic thesis that the order and connection of things is the same as the order and connection of ideas the unconscious wish behind his version of the Parmenidean thesis is clairvoyance. One thing that Freud said is particularly relevant in the present connection. He remarked that '. . . the philosophy of today has retained some essential features of the animistic mode of thought – the overvaluation of the magic of words and the belief that the real events in the world take the course which our thinking seeks to impose on them.'[47]

To return to the philosopher who claims to have demonstrated a contradiction in the meaning of the word 'motion' and who goes on to tell us that motion does not really occur, he does not say anything which is in conflict with the actual use of the word 'motion'. He has decided to contract the application of 'real' and withold its application from whatever 'is in motion' correctly applies to. He presents his semantic decision in the language of ontology and in this way creates the appearance of making a startling assertion about the true state of the cosmos. The argument he gives for his assertion makes it look as if by an analytical penetration into concepts or into the labyrinth of linguistic usage he was able to arrive at a basic truth about things.

Russell's remark that everyone except a philosopher can see the difference between a post and his idea of a post would seem to apply to Berkeley, whose view, taken literally, implies that a post is no more than an idea.[48] Samuel Johnson showed that he knew the difference between a thing and an idea of the thing by kicking a stone, but his action furnished no evidence against the philosophical view that to be a thing, such as a post or a stone, is to be perceived. One philosopher stated that our senses reveal only sense-data, or ideas, to us, but his inability to say what else our senses might reveal shows that he was not using language to make a statement of fact. Similarly, a Berkeleian who maintains that the existence of a thing depends upon its being perceived is unable to say what it would be like for an unperceived thing to exist, or what it would be like to encounter a thing which is not an idea. He is not using language to make a matter-of-fact statement about things, which is why Johnson's kicking a stone does not count as evidence against what he says. But thought of as intended to express an *a priori* proposition, the sentence 'To be a thing is to be perceived' would be equivalent to the entailment-sentence 'Being a thing *entails* being perceived'. We should then have to suppose that a Berkeleian has the idea that, as English is ordinarily used, the phrase 'unperceived thing' is devoid of conceptual content, and also that he has the idea that usage dictates the application of 'is perceived' to whatever 'thing' correctly applies to. Berkeley's advice to speak with the vulgar but to think with the learned shows that he had no such mistaken idea about usage. The proper conclusion to come to is that his view is nothing more substantial than a deceptively presented piece of changed terminology – a vacuously stretched use of 'perceived' which makes it applicable to whatever 'thing' correctly applies to.

The words 'It is impossible to think of what does not exist', which serve to reassure some philosophers and strike others as an extravagant violation of common sense, are of a kind with 'Motion does not really exist' and 'Things such as chimneys and mountains cannot exist unperceived'. They bring before us, in a veiled and dramatized form, an idle redistricting of terminology, i.e., a change in the application of an expression in dissociation from any practical intention to institute a change in its actual use. They present an academically contracted application of the expression 'thinks of what does not exist', an application whose range has been shrunk to zero. In his contrived way of speaking, in the special language game the philosopher plays, 'thinks of a thief who does not exist' has been shorn of its descriptive sense. In depriving the phrase of its descriptive use a philosopher also deprives the related antithetical phrase, 'thinks of a thief who exists', of its function to describe. But this he leaves unmentioned, and for good reason: his purpose is not to introduce practical changes in language, rather it is to create certain effects.

A Parmenidean metaphysician speaks the language of the learned without for a moment giving up the language of the vulgar. He asserts that one cannot think of what does not exist and that whatever is thought of must exist and also continues to speak, like everyone else, of things that do not exist as well as of things that do exist. Everyday language not only remains the instrument he uses to communicate factual information, it also serves as the constant backdrop in front of which he speaks his philosophical lines. The things he does with language would lose their power to bring into existence the illusion that he is expressing a theory about thoughts and things if, instead of being paraded in front of unchanged everyday language, they were *incorporated* into it. For then both expressions, 'thinks of a nonexisting thing' and 'thinks of an existing thing', would lose their use in sentences about states of affairs.

They would pass out of currency, and with their disappearance from language the philosophical theory would evaporate. In order to create an illusion with his altered nomenclature, ordinary, everyday language must remain intact, which is why philosophical talk has no tendency whatever to modify or in any way change ordinary talk. Wittgenstein explained what a philosopher is doing who says, 'Only my pain is real', as keeping ordinary language and putting another beside it,[49] a language in which 'his pain' has no application. Similarly, we may say that a philosopher who asserts 'It is impossible to think of things which once were and no longer are' is keeping ordinary language and putting another beside it, and that he does this for the magical effect he is able to produce.[50]

People have the idea not only that a technical philosopher advances theories about the nature and existence of things, but also that he arrives at them by some sort of controlled investigation. Freud thought that philosophy 'behaves like a science and works in part by the same methods'. And a leading contemporary philosopher has described philosophy as 'a wing of science where aspects of method are examined more deeply or in a wider perspective than elsewhere'.[51] Undoubtedly, philosophers have found reassurance and support in his words. But the idea that philosophy works like a science is as far removed from reality as is the idea that a philosophical view has theoretical content. A philosophical view is a semantic wind-egg, a bubble; and a semantic bubble is not the outcome of an investigation which employs scientific techniques or methods. If we can bring ourselves to scrutinize a philosophical argument with care, we shall find that what looks like a piece of scientific reasoning is of a kind with the theory it backs. A philosophical theory is constituted by an altered piece of language; and in general, an alteration in the range covered by an expression is bound up with likenesses and differences in the functioning of related expressions occurring in the statement of the argument. The wish to highlight a likeness between certain expressions or a difference between them requires introducing changes into the language which will do the work of accentuating or minimizing the likeness or difference. When these changes are presented in the ontological idiom they take the form of an argument for a theory. To illustrate, Heraclitus' statement, 'You cannot step into the same river twice' is a picturesque way of calling attention to the difference between the use of 'same' in 'same river' and its use in 'same street', coupled with the academic decision to discontinue applying 'same' to rivers and the like. Highlighting the point that there is only a difference of degree, not one of kind, between rivers and streets enables a philosopher to withhold applying the word 'same' not only to rivers but also to streets. The view that everything constantly changes, or that things are really events, is the outcome of these two moves with the word 'same'.

Moore had a poor opinion of the argument he formulated for the Bradleian-Parmenidean view; but in philosophy what is one man's poison is another man's meat, and what is a philosopher's poison at one time may well become his meat at another. Moore's formulation makes explicit a reason for holding the view. Regardless of whether it is thought bad or good it is a reason, which lends to the view the scientific air that it was arrived at by the employment of a method. Moore supposes Bradley to be arguing as follows: 'A thing cannot have a property, unless it is there to have it, and since unicorns. . .do have the property of being thought of, there certainly must be such things.' The implication of Moore's words is that a philosopher who says, 'It is impossible to think of what does not exist', has the idea that the phrase 'being thought of' denotes a property of things. According to this idea, to assert that the moon is being thought of is to assert that it has the property *being thought of*, just as to assert that the moon is round is to assert that it has the property *being round*. Moreover, *being thought of* is a property of such a kind that it can only be ascribed to a subject that is 'there to have it'. Wittgenstein's observation that you cannot hang a thief who does not exist but you can think of one who does not exist tells us the *sort* of property Moore takes a philosopher to understand by the expression 'being thought of': it denotes the sort of property he elsewhere calls a 'relational property'. In other words, it refers to a relation between objects, in the present connection, to an action that is performed on one thing by another. Moore's formulation of the Bradleian argument carries with it the suggestion that Bradley took the words 'thinks of the moon' as describing something that is being done to the moon. It suggests that Bradley, and others, have the idea that the use in the language of 'thinks of a thief' is like the use of 'hangs a thief', 'opens the door', and 'milks a goat'.

It is hardly necessary to remark that a philosopher gives no evidence of having this idea about the *actual* use of 'thinks of a thief'. Except for his philosophical theory and his philosophical argument he talks like everyone else. It is reasonable to suppose that instead of having a wrong idea about the use of 'thinks of' in expressions like 'thinks of a thief' and 'thinks of a unicorn' he wishes to assimilate it to transitive verbs occurring in such expressions as 'hangs a thief' and 'rides a unicorn'. When a philosopher insists that it is just as impossible to think of a unicorn that does not exist as it is to ride a unicorn that does not exist, he is not wrong about linguistic usage. Instead he is urging an artificial classification of 'thinks of' with transitive verbs denoting an action. A grammatical similarity appeals to him, and he accents it, while muting a semantic difference. The argument for the view that it is impossible to think of what does not exist is an academic grouping together of verbs which a grammatical likeness makes possible. The view itself is a vivid way of announcing this grouping.

The Parmenidean philosophical theory is a remarkable structure, one that has the substance of a cobweb and yet is as durable as the pyramids of Egypt. At the center of the structure is a piece of unconsciously reframed grammar that is presented in the nonverbal form of speech and projected onto the language in everyday use, without being incorporated into it. The result of this game with terminology is a lively illusion that a theory about the relation between thoughts and things is being presented and supported by a line of reasoning. Behind the verbal sleight of hand we can detect the archaic wish for effortless knowledge which is still active in the unconscious of many thinkers. There can be hardly any doubt that the overriding strength of this wish is responsible for the blindspots philosophers have to the anomalies of their discipline. Philosophers like Plato, Spinoza, Hegel, and Bradley subjectively picture themselves in the role of surveyor of all that there is and as composing the Cosmic Baedecker. The reality behind this gratifying image is less appealing: the spectator of the universe deflates into a juggler who plays empty tricks with grammar.

A number of things Wittgenstein has said carry with them the implication that a philosophical theory is a kind of neurotic symptom. Thus, he wrote, 'The philosopher is the man who has to cure himself of many sicknesses of the understanding. . .'[52] If indeed philosophy can be correctly described as a sickness, it is a sickness of which no philosopher wishes to be cured. His 'sickness' of the understanding, on which he looks as a lofty achievement, gives him pleasure and also has a commercial value. Seeing this, it is not hard to understand why a philosopher should be reluctant to open his eyes to the insubstantiality of his subject. The most powerful forces which work against his looking with analytical eye at his subject are hidden from him, one of which is the subjectively retained belief in the superhuman reach of his thought. This belief is of utmost importance to him, as it gives him the feeling that instead of being weak and at the mercy of external agencies he is powerful and self-sufficient. Linked with the idea of his omniscience is the belief in the causal power of thought, the power to think things into and out of existence and to have mental control over their behavior. The world systems of philosophers are not only the epic accounts of mental journeys, they also describe worlds that have been brought into existence by the minds of philosophers. In his unconscious a philosopher's mind has omniscience and omnipotence, and it is language which enables it to possess these attributes.

A philosophical theory is a complex structure the greater part of which is hidden in the unconscious. Its importance to the philosopher is great enough to prevent him from prying into its two uppermost parts, a gerrymandered piece of language and an intellectual illusion. He is like the king in the fairy tale whose exhibitionistic needs prevented him from seeing that the cloth which the two rascally tailors wove for his garment was transparent air. If we look into the Parmenidean thesis we can discern a further subterranean idea which is related to the fairy tale and may be useful to allude to briefly, as it will help us understand better the need of a philosopher to be deceived by his own production. For the philosopher is both tailor and king, deceiver and deceived. There are two equivalent formulations of the Parmenidean thesis,

> Whatever is thought of exists,
> It is impossible to think of what does not exist.

It is the second formulation which provides us with a clue as to what creates the need to be deceived. The suggestion of the words 'One cannot think of what does not exist' is that there is something whose nonexistence is *too painful* to be thought of, something whose nonexistence is 'unthinkable and unspeakable' and must be banished from the mind. It is not difficult to identify, with reasonable certainty, the anatomical entity that is referred to: it is not possessed by some and its loss is feared by others. In each case it is the unspeakable non-being, actual or feared. The importance to the philosopher of his sentence about the way thoughts and things are related is understandable. With it he is not only able to create an illusion which gives him narcissistic satisfaction, he is also able to fend off the invasion of anxiety.

NOTES

1 W.V. Quine, 'Philosophical Progress in Language Theory', *Metaphilosophy* 1, p. 2.
2 W.V. Quine, *Word and Object*, p. 207.
3 A.J. Ayer, *Russell and Moore: The Analytical Heritage*, p. 245.
4 *New Introductory Lectures on Psychoanalysis*, newly translated and edited by James Strachey, p. 16.
5 *Scientific Thought*, p. 19.
6 Ibid., p. 18.
7 Ibid., p. 18.
8 *Tractatus Logico-Philosophicus*, 4.112. Pears and McGuinness translation.
9 'Philosophy and "Common-Sense" ', in *G.E. Moore. Essays in Retrospect* (eds. Alice Ambrose and Morris Lazerowitz), p. 203.
10 'A Defence of Common Sense', in *Philosophical Papers*, p. 41.
11 James Strachey's literal translation is: 'With his nightcaps and the tatters of his dressing-gown he patches up the gaps in the structure of the universe.'
12 *New Introductory Lectures on Psychoanalysis*, pp. 160–61.
13 *Free Associations: Memories of a Psycho-analyst*, p. 60.
14 Ibid., p. 165.
15 Ibid., p. 15.
16 Ibid., p. 16.
17 This difference is dismissed by so-called conventionalists, but is developed into a theory about abstract objects by Platonists.
18 'A Reply to My Critics', in *The Philosophy of G.E. Moore*, The Library of Living Philosophers, Vol. IV (ed. by P.A. Schilpp), pp. 675–6.
19 A.J. Ayer, *Russell and Moore: The Analytical Heritage*, p. 180.
20 This is developed in detail in my *Philosophy and Illusion*, pp. 119–140.
21 *Objects of Thought*, by A.N. Prior. Reviewed in The Times Literary Supplement, 1971.
22 John Burnet, *Greek Philosophy*, p. 67.
23 *The Blue Book*, p. 31.
24 *The Principles of Mathematics*, p. 450.
25 Ibid., p. 451.
26 Ibid., pp. 450–61.
27 Ibid., p. 451.
28 Pp. 379–80.
29 p. 381.
30 *Philosophical Studies*, p. 215.
31 Ibid., p. 216.
32 Benjamin Farrington, *Greek Science. Its Meaning for Us*, p. 50.
33 For an extended discussion of Moore's disconcerting observation that philosophers hold views incompatible with what they know to be true, see 'Moore's Paradox' in my *The Structure of Metaphysics*.
34 For a discussion whether this is an actual or only a semantically contrived distinction, see my *Studies in Metaphilosophy*, p. 6.
35 *Some Main Problems of Philosophy*, Chapter I.
36 It is safe to say *all* philosophers: scratch the semantic veneer of the so-called linguistic philosopher and you will find the metaphysician underneath.

[37] *Remarks on the Foundations of Mathematics*, p. 37.

[38] *The Principles of Mathematics*, Introduction, p. vii.

[39] *Tractatus Logico-Philosophicus*, 4.461. Ogden translation.

[40] *An Enquiry Concerning Human Understanding*, Section XII, Part III.

[41] For an exposition of this thesis, see Norman Malcolm's 'Moore and Ordinary Language' in *The Philosophy of G.E. Moore*, The Library of Living Philosophers, Vol. IV (ed. P.S. Schilpp).

[42] *Appearance and Reality*, p. 43.

[43] *Philosophische Grammatik*, p. 462: Men are entangled in the net of language and do not know it. (My translation)

[44] *Philosophical Investigations*, p. 51.

[45] Wittgenstein has made the penetrating observation that 'Behind our thoughts, true or false, there is always to be found a dark background, which we are only later able to bring into the light and express as a thought' (*Notebooks 1914–1916*, p. 36e). There can be no doubt that the dark background contains unconscious material.

[46] One thing which indicates a weakened sense of reality with respect to his discipline is the philosopher's unconcern about the total absence of stable results in it. Occasionally in its busy history, now covering a period of twenty-five centuries, a philosopher has awakened to the ubiquitous chaos that reigns in his subject, but his return to peaceful slumber has always been prompt.

[47] 'The Question of a *Weltanschauung*', Lecture XXXV, p. 166, in *New Introductory Lectures on Psychoanalysis*.

[48] More accurately, a system of ideas in the mind of God. Phenomenalism, which comes from this view, is that a physical object is a system of actual and *possible* sense data.

[49] The Yellow Book.

[50] I do not wish to imply that Wittgenstein would go on to this sort of conclusion.

[51] W.V. Quine, 'Philosophical Progress in Language Theory', op. cit., p. 2.

[52] *Remarks on the Foundations of Mathematics*, p. 157. He described his own treatment of a philosophical question as being 'like the treatment of an illness' (*Philosophical Investigations*, p. 91). It is interesting to realize that despite his rejecting psychoanalysis (after first having admired Freud's ideas) Wittgenstein at times took, in external respects, a psychoanalytical approach to philosophy. What might be called a displaced return of the rejected appears to have taken place in him.

BIBLIOGRAPHY OF WORKS CITED

Ambrose, Alice, *Essays in Analysis*, Allen and Unwin, London, 1966.

Aristotle, *Metaphysics*. Book Z, Clarendon Press, Oxford, second edition, 1928. Translated by W.D. Ross.

Austin, J.L., *Philosophical Papers*, Oxford University Press, London, 1969. Edited by J.O. Urmson and G.J. Warnock.

Ayer, A.J., *Language, Truth and Logic*, Victor Gollancz, London, revised edition, 1951, (originally published 1936).

Ayer, A.J., *Philosophical Essays*, Macmillan & Co., London, 1954.

Ayer, A.J., *The Problem of Knowledge*, Penguin Books, Harmondsworth, 1956.

Ayer, A.J., *Russell and Moore: The Analytical Heritage*, Harvard University Press, Cambridge, Mass., 1971.

Bell, E.T., *Men of Mathematics*, Simon and Schuster, New York, 1937.

Blanshard, Brand, 'In Defense of Metaphysics', in *Metaphysics. Readings and Reappraisals*, Prentice-Hall, Englewood Cliffs, 1966. Edited by William E. Kennick and Morris Lazerowitz.

Blanshard, Brand, Review of *Philosophy and Illusion, Metaphilosophy*, 1, 1970.

Bradley, F.H., *Appearance and Reality*, Allen & Unwin, London, 7th Impression, 1920.

Broad, C.D., 'Philosophy and "Common-Sense" ', in *G.E. Moore Essays in Retrospect*, Allen & Unwin, London, 1970. Edited by Alice Ambrose and Morrise Lazerowitz.

Broad, C.D., *Scientific Thought*, Harcourt Brace, N.Y., 1927.

Burnet, John, *Greek Philosophy, Pt. I, Thales to Plato*, Macmillan, London, 1928.

Drennen, R., *A Modern Introduction to Metaphysics; Readings from Classical and Contemporary Sources*, Free Press of Glencoe, N.Y., 1962.

Farrington, Benjamin, *Greek Science, Its Meaning for Us*, Penguin Books, Harmondsworth, 1944.

Feuer, Lewis Samuel, *Spinoza and the Rise of Liberalism*, Beacon Press, Boston, 1958.

Freud, Sigmund, 'The Antithetical Sense of Primal Words', *Collected Papers*, Vol. IV, International Psychoanalytical Library, Hogarth Press and The Institute of Psychoanalysis, London, 1925. Edited by Ernest Jones.

Freud, Sigmund, *An Autobiographical Study*, International Psycho-analytical Library, No. 26, Hogarth Press and The Institute of Psycho-analysis, London, second edition, 1946. Translated by James Strachey; edited by Ernest Jones. Standard Edition of the Complete Psychological Works, Vol. 20.

Freud, Sigmund, *A General Introduction to Psychoanalysis*, Garden City Publishing Co., Garden City, N.Y., 1938. Translated by Joan Riviere. Standard Edition of the Complete Psychological Works, Vol. 15.

Freud, Sigmund, *New Introductory Lectures on Psychoanalysis*, W.W. Norton, N.Y., 1965. Newly Translated and edited by James Strachey. (First published, 1933). Standard Edition of the Complete Psychological Works, Vol. 22.

Gamow, George, *One Two Three. . .Infinity*, Viking Press, N.Y., 1949.

Gellner, Ernest, *Words and Things*, Victor Gollancz, London, 1959.

Gomperz, Theodor, *The Greek Thinkers*, Vol. I, John Murray, London, reprinted, 1949. Translated by Magnus Laurie.

Hahn, Hans, 'Infinity', in *The World of Mathematics*, Vol. 3, Simon and Schuster, New York, 1956. Edited by James R. Newman.

Hanly, Charles, 'Wittgenstein and Psychoanalysis', in *Ludwig Wittgenstein: Philosophy and Language*, Allen and Unwin, London, 1972. Edited by Alice Ambrose and Morris Lazerowitz.

Hanly, Charles and Lazerowitz, Morris (ed.), *Psychoanalysis and Philosophy*, International Universities Press, N.Y., 1970.

Herrmann, Paul, *Conquest by Man*, Harper Brothers, N.Y., 1954. Translated by Michael Bullock. Published in Germany under the title: *Sieben Vorbei und Acht Verweht.*

Heyting, A., *Intuitionism. An Introduction*, North-Holland Publishing Co., Amsterdam, 1956.

Hook, Sidney, (ed.), *Psychoanalysis, Scientific Method, and Philosophy*, second New York University Institute of Philosophy, New York University Press, N.Y., 1959.

Hume, David, *An Enquiry Concerning Human Understanding*, The Clarendon Press, Oxford, second edition, 1902. Edited by L.A. Selby-Bigge.

Hume, David, *A Treatise of Human Nature*, Clarendon Press, Oxford, first edition, 1888, reprinted 1941. Edited by L.A. Selby-Bigge.

James, William, *The Varieties of Religious Experience*, Longmans, Green, & Co., 35th impression, N.Y., 1925.

Jones, Ernest, *Free Associations: Memories of a Psycho-analyst*, Basic Books, N.Y., 1959.

Jourdain, P.E.B., *The Philosophy of Mr B*RTR*ND R*SS*LL*, Allen and Unwin, London, 1918.

Kant, Immanuel, *Critique of Pure Reason*, Macmillan, London, second impression, 1933. Translated by Norman Kemp Smith.

Kant, Immanuel, *Prolegomena to Any Future Metaphysics*, The Library of Liberal Arts, No. 27, N.Y., 1950. The Maffy-Carus translation, revised by Lewis White Beck.

Lazerowitz, Morris, *Note*, in *Metaphilosophy*, **1**, 1970.

Lazerowitz, Morris, *Philosophy and Illusion*, Allen and Unwin, London, 1968.

Lazerowitz, Morris, *The Structure of Metaphysics*, Routledge and Kegan Paul, London, 1955.

Lazerowitz, Morris, *Studies in Metaphilosophy*, Routledge and Kegan Paul, London, 1964.

Lazerowitz, Morris and Kennick, W.E. (ed.), *Metaphysics. Readings and Reappraisals*, Prentice-Hall, Englewood Cliffs, 1966.

Lazerowitz, Morris and Ambrose, Alice (ed.), *Ludwig Wittgenstein. Philosophy and Language*, Allen and Unwin, London. 1972.

Lazerowitz, Morris and Ambrose, Alice (ed.), *G.E. Moore. Essays in Retrospect*, Allen and Unwin, London, 1970.

Lazerowitz, Morris and Hanly, Charles (ed.), *Psychoanalysis and Philosophy*, International Universities Press, 1970.

Leibniz, G.W., *Philosophische Schriften*, Vol. I, Gerhardt's edition, Berlin, 1875–90.
Lewis, C.I., *An Analysis of Knowledge and Valuation*, Open Court, LaSalle, 1946.
Locke, John, *An Essay Concerning Human Understanding*, Clarendon Press, Oxford, Impression of 1928. Abridged and edited by A.S. Pringle-Pattison.
Malcolm, Norman, *Knowledge and Certainty*, Prentice-Hall, Englewood Cliffs, 1963.
Malcolm, Norman, *Ludwig Wittgenstein. A Memoir*, Oxford University Press, London, 1958.
Malcolm, Norman, 'Moore and Ordinary Language', in *The Philosophy of G.E. Moore*, The Library of Living Philosophers, Vol. IV, Northwestern University, Evanston and Chicago, 1942. Edited by P.A. Schilpp.
Mill, J.S., *A System of Logic*, Harper and Brothers, N.Y., 1856.
Moore, G.E., 'A Defence of Common Sense', in *Philosophical Papers*, Allen and Unwin, London, 1959. First published in *Contemporary British Philosophy*, Second Series, Allen and Unwin, London, 1925, edited by J.H. Muirhead.
Moore, G.E., *Ethics*, Home University Library of Modern Knowledge, No. 52, Williams and Norgate, London.
Moore, G.E., *Philosophical Papers*, Allen and Unwin, London, 1959.
Moore, G.E., *Philosophical Studies*, Routledge and Kegan Paul, London, 1922.
Moore, G.E., *Principia Ethica*, Cambridge University Press, Cambridge, 1922.
Moore, G.E., 'A Reply to My Critics', in *The Philosophy of G.E. Moore*, Library of Living Philosophers, Vol. IV, Northwestern University, Evanston and Chicago, 1942. Edited by P.A. Schilpp.
Moore, G.E., *Some Main Problems of Philosophy*, Allen & Unwin, London, 1953.
Plato, *Phaedrus*, Random House, N.Y., 1937. Translated by B. Jowett from the third edition of *The Dialogues of Plato*.
Quine, W.V., *From a Logical Point of View*, Harper Torchbooks, Harper and Row, N.Y., second revised edition, 1963.
Quine, W.V., 'Philosophical Progress in Language Theory', *Metaphilosophy* 1, 1970.
Quine, W.V., 'Semantics and Abstract Objects', *Proc. American Academy of Sciences* 80, 1951.
Quine, W.V., *Word and Object*, Wiley and Sons, N.Y., 1960.
Ramsey, F.P., *The Foundations of Mathematics*, Harcourt, Brace, & Co., N.Y., 1931. Edited by R.B. Braithwaite, with a Preface by G.E. Moore.
Raphael, D.D., 'A Critical Study. *The Structure of Metaphysics* by Morris Lazerowitz', *The Philosophical Quarterly* 7, 1957.
Rhees, Rush, 'Conversations on Freud', *Ludwig Wittgenstein, Lectures and Conversations on Aesthetics, Psychology and Religious Belief*, University of California Press, Berkeley, 1967. Edited by Cyril Barrett.
Russell, Bertrand, Introduction to *Words and Things*, by Ernest Gellner, Victor Gollancz, London, 1959.
Russell, Bertrand, 'The Limits of Empiricism', *Proc. Aristotelian Society* 36, 1936.
Russell, Bertrand, *Mysticism and Logic*, Allen and Unwin, London, 9th edition, 1950.
Russell, Bertrand, *Our Knowledge of the External World*, Open Court, Chicago, 1914.
Russell, Bertrand, *An Introduction to Mathematical Philosophy*, Allen and Unwin, London, second edition, 1920.
Russell, Bertrand, *The Principles of Mathematics*, W.W. Norton, N.Y., second edition, 1938.

Russell, Bertrand and Whitehead, A.N., *Principia Mathematica*, Vol. I, Cambridge University Press, Cambridge, second edition, 1925.

Schilpp, P.A. (ed.), *The Philosophy of G.E. Moore*, Library of Living Philosophers, Vol. IV, Northwestern University, Evanston and Chicago, 1942.

Spinoza, Benedict de, *Ethics*, Wiley, N.Y., 1901. Translated by R.H.M. Elwes.

Stace, W.T., 'Mysticism and Human Reason', Reicker Memorial Lecture, No. 1, *University of Arizona Bulletin Series*, 1955.

Stace, W.T., *Mysticism and Philosophy*, Lippincott, N.Y., 1960.

Sterba, Richard, 'Remarks on Mystic States', *American Imago* **25**.

Times Literary Supplement review of A.N. Prior's *Objects of Thought*, 1971.

von Wright, G.H., 'Biographical Sketch', in *Ludwig Wittgenstein: A Memoir*, by Norman Malcolm. Oxford University Press, London, 1958.

von Wright, G.H., 'Deontic Logic and the Theory of Conditions', *Critica* 2, 1968.

Warkins, J.W.N., 'Confirmable and Influential Metaphysics', *Mind* 6, 1958.

Watkins, J.W.N., 'Word Magic and the Trivialization of Philosophy', *Ratio* 7, 1965.

Wilder, R.L., *Introduction to the Foundations of Mathematics*, John Wiley and Sons, second edition, 1965.

Wisdom, John, *Philosophy and Psychoanalysis*, Basil Blackwell, Oxford, 1953.

Wisdom, J.O., *The Unconscious Origins of Berkeley's Philosophy*, The Hogarth Press and The Institute of Psycho-analysis, London, 1953.

Wittgenstein, Ludwig, *The Blue and Brown Books, Preliminary Studies for the "Philosophical Investigations"*, Basil Blackwell, Oxford, 1958.

Wittgenstein, Ludwig, *On Certainty*, Basil Blackwell, Oxford, 1969. Edited by G.E.M. Anscombe and G.H. von Wright. Translated by Denis Paul and G.E.M. Anscombe.

Wittgenstein, Ludwig, *Notebooks 1914–1916*, Basil Blackwell, Oxford, 1969.

Wittgenstein, Ludwig, *Philosophical Investigations*, Basil Blackwell, Oxford, 1953. Translated by G.E.M. Anscombe.

Wittgenstein, Ludwig, *Philosophische Grammatik*, Schriften 4, Suhrkamp Frankfurt am Main, 1969. Edited by Rush Rhees.

Wittgenstein, Ludwig, *Remarks on the Foundations of Mathematics*, Macmillan, N.Y., 1956. Edited by G.H. von Wright, Rush Rhees, and G.E. Anscombe. Translated by G.E.M. Anscombe.

Wittgenstein, Ludwig, *Tractatus Logico-Philosophicus*, Kegan Paul, Trench, Trubner and Co., London, 1922. Translated by C.K. Ogden.

Wittgenstein, Ludwig, *Tractatus Logico-Philosophicus*, Routledge and Kegan Paul, London, 1961. Translated by D.F. Pears and B.F. McGuinness.

Wittgenstein, Ludwig, unpublished notes, called The Yellow Book, taken by Alice Ambrose and Margaret Masterman, 1933–34, in the intervals between dictation of *The Blue Book*.

Wolf, A. (ed.), *The Oldest Biography of Spinoza*, London, 1927.

INDEX

SYNTHESE LIBRARY

Monographs on Epistemology, Logic, Methodology,
Philosophy of Science, Sociology of Science and of Knowledge, and on the
Mathematical Methods of Social and Behavioral Sciences

Managing Editor:
JAAKKO HINTIKKA (Academy of Finland and Stanford University)

Editors:

ROBERT S. COHEN (Boston University)
DONALD DAVIDSON (University of Chicago)
GABRIËL NUCHELMANS (University of Leyden)
WESLEY C. SALMON (University of Arizona)

1. J. M. Bocheński, *A Precis of Mathematical Logic.* 1959, X + 100 pp.
2. P. I. Guiraud, *Problèmes et méthodes de la statistique linguistique.* 1960, VI + 146 pp.
3. Hans Freudenthal (ed.), *The Concept and the Role of the Model in Mathematics and Natural and Social Sciences, Proceedings of a Colloquium held at Utrecht, The Netherlands, January 1960.* 1961, VI + 194 pp.
4. Evert W. Beth, *Formal Methods. An Introduction to Symbolic Logic and the Study of Effective Operations in Arithmetic and Logic.* 1962, XIV + 170 pp.
5. B. H. Kazemier and D. Vuysje (eds.), *Logic and Language. Studies Dedicated to Professor Rudolf Carnap on the Occasion of His Seventieth Birthday.* 1962, VI + 256 pp.
6. Marx W. Wartofsky (ed.), *Proceedings of the Boston Colloquium for the Philosophy of Science, 1961-1962,* Boston Studies in the Philosophy of Science (ed. by Robert S. Cohen and Marx W. Wartofsky), Volume I. 1973, VIII + 212 pp.
7. A. A. Zinov'ev, *Philosophical Problems of Many-Valued Logic.* 1963, XIV + 155 pp.
8. Georges Gurvitch, *The Spectrum of Social Time.* 1964, XXVI + 152 pp.
9. Paul Lorenzen, *Formal Logic.* 1965, VIII + 123 pp.
10. Robert S. Cohen and Marx W. Wartofsky (eds.), *In Honor of Philipp Frank,* Boston Studies in the Philosophy of Science (ed. by Robert S. Cohen and Marx W. Wartofsky), Volume II. 1965, XXXIV + 475 pp.
11. Evert W. Beth, *Mathematical Thought. An Introduction to the Philosophy of Mathematics.* 1965, XII + 208 pp.
12. Evert W. Beth and Jean Piaget, *Mathematical Epistemology and Psychology.* 1966, XII + 326 pp.
13. Guido Küng, *Ontology and the Logistic Analysis of Language. An Enquiry into the Contemporary Views on Universals.* 1967, XI + 210 pp.
14. Robert S. Cohen and Marx W. Wartofsky (eds.), *Proceedings of the Boston Colloquium for the Philosophy of Science 1964-1966, in Memory of Norwood Russell Hanson,* Boston Studies in the Philosophy of Science (ed. by Robert S. Cohen and Marx W. Wartofsky), Volume III. 1967, XLIX + 489 pp.

15. C. D. Broad, *Induction, Probability, and Causation. Selected Papers.* 1968, XI + 296 pp.
16. Günther Patzig, *Aristotle's Theory of the Syllogism. A Logical-Philosophical Study of Book A of the Prior Analytics.* 1968, XVII + 215 pp.
17. Nicholas Rescher, *Topics in Philosophical Logic.* 1968, XIV + 347 pp.
18. Robert S. Cohen and Marx W. Wartofsky (eds.), *Proceedings of the Boston Colloquium for the Philosophy of Science 1966-1968,* Boston Studies in the Philosophy of Science (ed. by Robert S. Cohen and Marx W. Wartofsky), Volume IV. 1969, VIII + 537 pp.
19. Robert S. Cohen and Marx W. Wartofsky (eds.), *Proceedings of the Boston Colloquium for the Philosophy of Science 1966-1968,* Boston Studies in the Philosophy of Science (ed. by Robert S. Cohen and Marx W. Wartofsky), Volume V. 1969, VIII + 482 pp.
20. J.W. Davis, D. J. Hockney, and W. K. Wilson (eds.), *Philosophical Logic.* 1969, VIII + 277 pp.
21. D. Davidson and J. Hintikka (eds.), *Words and Objections: Essays on the Work of W. V. Quine.* 1969, VIII + 366 pp.
22. Patrick Suppes, *Studies in the Methodology and Foundations of Science. Selected Papers from 1911 to 1969.* 1969, XII + 473 pp.
23. Jaakko Hintikka, *Models for Modalities. Selected Essays.* 1969, IX + 220 pp.
24. Nicholas Rescher *et al.* (eds.), *Essays in Honor of Carl G. Hempel. A Tribute on the Occasion of His Sixty-Fifth Birthday.* 1969, VII + 272 pp.
25. P. V. Tavanec (ed.), *Problems of the Logic of Scientific Knowledge.* 1969, XII + 429 pp.
26. Marshall Swain (ed.), *Induction, Acceptance, and Rational Belief.* 1970, VII + 232 pp.
27. Robert S. Cohen and Raymond J. Seeger (eds.), *Ernst Mach: Physicist and Philosopher,* Boston Studies in the Philosophy of Science (ed. by Robert S. Cohen and Marx W. Wartofsky), Volume VI. 1970, VIII + 295 pp.
28. Jaakko Hintikka and Patrick Suppes, *Information and Inference.* 1970, X + 336 pp.
29. Karel Lambert, *Philosophical Problems in Logic. Some Recent Developments.* 1970, VII + 176 pp.
30. Rolf A. Eberle, *Nominalistic Systems.* 1970, IX + 217 pp.
31. Paul Weingartner and Gerhard Zecha (eds.), *Induction, Physics, and Ethics: Proceedings and Discussions of the 1968 Salzburg Colloquium in the Philosophy of Science.* 1970, X + 382 pp.
32. Evert W. Beth, *Aspects of Modern Logic.* 1970, XI + 176 pp.
33. Risto Hilpinen (ed.), *Deontic Logic: Introductory and Systematic Readings.* 1971, VII + 182 pp.
34. Jean-Louis Krivine, *Introduction to Axiomatic Set Theory.* 1971, VII + 98 pp.
35. Joseph D. Sneed, *The Logical Structure of Mathematical Physics.* 1971, XV + 311 pp.
36. Carl R. Kordig, *The Justification of Scientific Change.* 1971, XIV + 119 pp.
37. Milič Čapek, *Bergson and Modern Physics,* Boston Studies in the Philosophy of Science (ed. by Robert S. Cohen and Marx W. Wartofsky), Volume VII. 1971, XV + 414 pp.

38. Norwood Russell Hanson, *What I Do Not Believe, and Other Essays* (ed. by Stephen Toulmin and Harry Woolf). 1971, XII + 390 pp.
39. Roger C. Buck and Robert S. Cohen (eds.), *PSA 1970. In Memory of Rudolf Carnap*, Boston Studies in the Philosophy of Science (ed. by Robert S. Cohen and Marx W. Wartofsky), Volume VIII. 1971, LXVI + 615 pp. Also available as paperback.
40. Donald Davidson and Gilbert Harman (eds.), *Semantics of Natural Language*. 1972, X + 769 pp. Also available as paperback.
41. Yehoshua Bar-Hillel (ed.), *Pragmatics of Natural Languages*. 1971, VII + 231 pp.
42. Sören Stenlund, *Combinators, λ-Terms and Proof Theory*. 1972, 184 pp.
43. Martin Strauss, *Modern Physics and Its Philosophy. Selected Papers in the Logic, History, and Philosophy of Science*. 1972, X + 297 pp.
44. Mario Bunge, *Method, Model and Matter*. 1973, VII + 196 pp.
45. Mario Bunge, *Philosophy of Physics*. 1973, IX + 248 pp.
46. A. A. Zinov'ev, *Foundations of the Logical Theory of Scientific Knowledge (Complex Logic)*, Boston Studies in the Philosophy of Science (ed. by Robert S. Cohen and Marx W. Wartofsky), Volume IX. Revised and enlarged English edition with an appendix, by G. A. Smirnov, E. A. Sidorenka, A. M. Fedina, and L. A. Bobrova. 1973, XXII + 301 pp. Also available as paperback
47. Ladislav Tondl, *Scientific Procedures*, Boston Studies in the Philosophy of Science (ed. by Robert S. Cohen and Marx W. Wartofsky), Volume X. 1973, XII + 268 pp. Also available as paperback.
48. Norwood Russell Hanson, *Constellations and Conjectures* (ed. by Willard C. Humphreys, Jr.). 1973, X + 282 pp.
49. K. J. J. Hintikka, J. M. E. Moravcsik, and P. Suppes (eds.), *Approaches to Natural Language. Proceedings of the 1970 Stanford Workshop on Grammar and Semantics*, 1973, VIII + 526 pp. Also available as paperback.
50. Mario Bunge (ed.), *Exact Philosophy − Problems, Tools, and Goals*. 1973, X + 214 pp.
51. Radu J. Bogdan and Ilkka Niiniluoto (eds.), *Logic, Language, and Probability. A Selection of Papers Contributed to Sections IV, VI, and XI of the Fourth International Congress for Logic, Methodology, and Philosophy of Science, Bucharest, September 1971*. 1973, X + 323 pp.
52. Glenn Pearce and Patrick Maynard (eds.), *Conceptual Chance*. 1973, XII + 282 pp.
53. Ilkka Niiniluoto and Raimo Tuomela, *Theoretical Concepts and Hypothetico-Inductive Inference*. 1973, VII + 264 pp.
54. Roland Fraïssé, *Course of Mathematical Logic − Volume 1: Relation and Logical Formula*. 1973, XVI + 186 pp. Also available as paperback.
55. Adolf Grünbaum, *Philosophical Problems of Space and Time*. Second, enlarged edition, Boston Studies in the Philosophy of Science (ed. by Robert S. Cohen and Marx W. Wartofsky), Volume XII. 1973, XXIII + 884 pp. Also available as paperback.
56. Patrick Suppes (ed.), *Space, Time, and Geometry*. 1973, XI + 424 pp.
57. Hans Kelsen, *Essays in Legal and Moral Philosophy*, selected and introduced by Ota Weinberger. 1973, XXVIII + 300 pp.
58. R. J. Seeger and Robert S. Cohen (eds.), *Philosophical Foundations of Science. Proceedings of an AAAS Program, 1969*, Boston Studies in the Philosophy of

Science (ed. by Robert S. Cohen and Marx W. Wartofsky), Volume XI. 1974, X + 545 pp. Also available as paperback.

59. Robert S. Cohen and Marx W. Wartofsky (eds.), *Logical and Epistemological Studies in Contemporary Physics,* Boston Studies in the Philosophy of Science (ed. by Robert S. Cohen and Marx W. Wartofsky), Volume XIII. 1973, VIII + 462 pp. Also available as paperback.

60. Robert S. Cohen and Marx W. Wartofsky (eds.), *Methodological and Historical Essays in the Natural and Social Sciences. Proceedings of the Boston Colloquium for the Philosophy of Science, 1969-1972,* Boston Studies in the Philosophy of Science (ed. by Robert S. Cohen and Marx W. Wartofsky), Volume XIV. 1974, VIII + 405 pp. Also available as paperback.

61. Robert S. Cohen, J. J. Stachel and Marx W. Wartofsky (eds.), *For Dirk Struik. Scientific, Historical and Political Essays in Honor of Dirk J. Struik,* Boston Studies in the Philosophy of Science (ed. by Robert S. Cohen and Marx W. Wartofsky), Volume XV. 1974, XXVII + 652 pp. Also available as paperback.

62. Kazimierz Ajdukiewicz, *Pragmatic Logic,* transl. from the Polish by Olgierd Wojtasiewicz. 1974, XV + 460 pp.

63. Sören Stenlund (ed.), *Logical Theory and Semantic Analysis. Essays Dedicated to Stig Kanger on His Fiftieth Birthday.* 1974, V + 217 pp.

64. Kenneth F. Schaffner and Robert S. Cohen (eds.), *Proceedings of the 1972 Biennial Meeting, Philosophy of Science Association,* Boston Studies in the Philosophy of Science (ed. by Robert S. Cohen and Marx W. Wartofsky), Volume XX. 1974, IX + 444 pp. Also available as paperback.

65. Henry E. Kyburg, Jr., *The Logical Foundations of Statistical Inference.* 1974, IX + 421 pp.

66. Marjorie Grene, *The Understanding of Nature: Essays in the Philosophy of Biology,* Boston Studies in the Philosophy of Science (ed. by Robert S. Cohen and Marx W. Wartofsky), Volume XXIII. 1974, XII + 360 pp. Also available as paperback.

67. Jan M. Broekman, *Structuralism: Moscow, Prague, Paris.* 1974, IX + 117 pp.

68. Norman Geschwind, *Selected Papers on Language and the Brain,* Boston Studies in the Philosophy of Science (ed. by Robert S. Cohen and Marx W. Wartofsky), Volume XVI. 1974, XII + 549 pp. Also available as paperback.

69. Roland Fraïssé, *Course of Mathematical Logic* – Volume 2: *Model Theory.* 1974, XIX + 192 pp.

70. Andrzej Grzegorczyk, *An Outline of Mathematical Logic. Fundamental Results and Notions Explained with All Details.* 1974, X + 596 pp.

71. Franz von Kutschera, *Philosophy of Language.* 1975, VII + 305 pp.

72. Juha Manninen and Raimo Tuomela (eds.), *Essays on Explanation and Understanding. Studies in the Foundations of Humanities and Social Sciences.* 1976, VII + 440 pp.

73. Jaakko Hintikka (ed.), *Rudolf Carnap, Logical Empiricist. Materials and Perspectives.* 1975, LXVIII + 400 pp.

74. Milič Čapek (ed.), *The Concepts of Space and Time. Their Structure and Their Development,* Boston Studies in the Philosophy of Science (ed. by Robert S. Cohen and Marx W. Wartofsky), Volume XXII. 1976, LVI + 570 pp. Also available as paperback.

75. Jaakko Hintikka and Unto Remes, *The Method of Analysis. Its Geometrical Origin and Its General Significance,* Boston Studies in the Philosophy of Science (ed. by Robert S. Cohen and Marx W. Wartofsky), Volume XXV. 1974, XVIII + 144 pp. Also available as paperback.

76. John Emery Murdoch and Edith Dudley Sylla, *The Cultural Context of Medieval Learning. Proceedings of the First International Colloquium on Philosophy, Science, and Theology in the Middle Ages – September 1973,* Boston Studies in the Philosophy of Science (ed. by Robert S. Cohen and Marx W. Wartofsky), Volume XXVI. 1975, X + 566 pp. Also available as paperback.

77. Stefan Amsterdamski, *Between Experience and Metaphysics. Philosophical Problems of the Evolution of Science,* Boston Studies in the Philosophy of Science (ed. by Robert S. Cohen and Marx W. Wartofsky), Volume XXXV. 1975, XVIII + 193 pp. Also available as paperback.

78. Patrick Suppes (ed.), *Logic and Probability in Quantum Mechanics.* 1976, XV + 541 pp.

79. H. von Helmholtz, *Epistemological Writings.* (A New Selection Based upon the 1921 Volume edited by Paul Hertz and Moritz Schlick, Newly Translated and Edited by R. S. Cohen and Y. Elkana), Boston Studies in the Philosophy of Science, Volume XXXVII. 1977 (forthcoming).

80. Joseph Agassi, *Science in Flux,* Boston Studies in the Philosophy of Science (ed. by Robert S. Cohen and Marx W. Wartofsky), Volume XXVIII. 1975, XXVI + 553 pp. Also available as paperback.

81. Sandra G. Harding (ed.), *Can Theories Be Refuted? Essays on the Duhem-Quine Thesis.* 1976, XXI + 318 pp. Also available as paperback.

82. Stefan Nowak, *Methodology of Sociological Research: General Problems.* 1977, XVIII + 504 pp. (forthcoming).

83. Jean Piaget, Jean-Blaise Grize, Alina Szeminska, and Vinh Bang, *Epistemology and Psychology of Functions.* 1977 (forthcoming).

84. Marjorie Grene and Everett Mendelsohn (eds.), *Topics in the Philosophy of Biology,* Boston Studies in the Philosophy of Science (ed. by Robert S. Cohen and Marx W. Wartofsky), Volume XXVII. 1976, XIII + 454 pp. Also available as paperback.

85. E. Fischbein, *The Intuitive Sources of Probabilistic Thinking in Children.* 1975, XIII + 204 pp.

86. Ernest W. Adams, *The Logic of Conditionals. An Application of Probability to Deductive Logic.* 1975, XIII + 156 pp.

87. Marian Przełęcki and Ryszard Wójcicki (eds.), *Twenty-Five Years of Logical Methodology in Poland.* 1977, VIII + 803 pp. (forthcoming).

88. J. Topolski, *The Methodology of History.* 1976, X + 673 pp.

89. A. Kasher (ed.), *Language in Focus: Foundations, Methods and Systems. Essays Dedicated to Yehoshua Bar-Hillel,* Boston Studies in the Philosophy of Science (ed. by Robert S. Cohen and Marx W. Wartofsky), Volume XLIII. 1976, XXVIII + 679 pp. Also available as paperback.

90. Jaakko Hintikka, *The Intentions of Intentionality and Other New Models for Modalities.* 1975, XVIII + 262 pp. Also available as paperback.

91. Wolfgang Stegmüller, *Collected Papers on Epistemology, Philosophy of Science and History of Philosophy,* 2 Volumes, 1977 (forthcoming).

92. Dov M. Gabbay, *Investigations in Modal and Tense Logics with Applications to Problems in Philosophy and Linguistics.* 1976, XI + 306 pp.
93. Radu J. Bogdan, *Local Induction.* 1976, XIV + 340 pp.
94. Stefan Nowak, *Understanding and Prediction: Essays in the Methodology of Social and Behavioral Theories.* 1976, XIX + 482 pp.
95. Peter Mittelstaedt, *Philosophical Problems of Modern Physics,* Boston Studies in the Philosophy of Science (ed. by Robert S. Cohen and Marx W. Wartofsky), Volume XVIII. 1976, X + 211 pp. Also available as paperback.
96. Gerald Holton and William Blanpied (eds.), *Science and Its Public: The Changing Relationship,* Boston Studies in the Philosophy of Science (ed. by Robert S. Cohen and Marx W. Wartofsky), Volume XXXIII. 1976, XXV + 289 pp. Also available as paperback.
97. Myles Brand and Douglas Walton (eds.), *Action Theory. Proceedings of the Winnipeg Conference on Human Action, Held at Winnipeg, Manitoba, Canada, 9-11 May 1975.* 1976, VI + 345 pp.
98. Risto Hilpinen, *Knowledge and Rational Belief.* 1978 (forthcoming).
99. R. S. Cohen, P. K. Feyerabend, and M. W. Wartofsky (eds.), *Essays in Memory of Imre Lakatos,* Boston Studies in the Philosophy of Science (ed. by Robert S. Cohen and Marx W. Wartofsky), Volume XXXIX. 1976, XI + 762 pp. Also available as paperback.
100. R. S. Cohen and J. Stachel (eds.), *Leon Rosenfeld, Selected Papers.* Boston Studies in the Philosophy of Science (ed. by Robert S. Cohen and Marx W. Wartofsky), Volume XXI. 1977 (forthcoming).
101. R. S. Cohen, C. A. Hooker, A. C. Michalos, and J. W. van Evra (eds.), *PSA 1974: Proceedings of the 1974 Biennial Meeting of the Philosophy of Science Association,* Boston Studies in the Philosophy of Science (ed. by Robert S. Cohen and Marx W. Wartofsky), Volume XXXII. 1976, XIII + 734 pp. Also available as paperback.
102. Yehuda Fried and Joseph Agassi, *Paranoia: A Study in Diagnosis,* Boston Studies in the Philosophy of Science (ed. by Robert S. Cohen and Marx W. Wartofsky), Volume L. 1976, XV + 212 pp. Also available as paperback.
103. Marian Przełęcki, Klemens Szaniawski, and Ryszard Wójcicki (eds.), *Formal Methods in the Methodology of Empirical Sciences.* 1976, 455 pp.
104. John M. Vickers, *Belief and Probability.* 1976, VIII + 202 pp.
105. Kurt H. Wolff, *Surrender and Catch: Experience and Inquiry Today,* Boston Studies in the Philosophy of Science (ed. by Robert S. Cohen and Marx W. Wartofsky), Volume LI. 1976, XII + 410 pp. Also available as paperback.
106. Karel Kosík, *Dialectics of the Concrete,* Boston Studies in the Philosophy of Science (ed. by Robert S. Cohen and Marx W. Wartofsky), Volume LII. 1976, VIII + 158 pp. Also available as paperback.
107. Nelson Goodman, *The Structure of Appearance,* Boston Studies in the Philosophy of Science (ed. by Robert S. Cohen and Marx W. Wartofsky), Volume LIII. 1977 (forthcoming).
108. Jerzy Giedymin (ed.), *Kazimierz Ajdukiewicz: Scientific World-Perspective and Other Essays, 1931–1963.* 1977 (forthcoming).
109. Robert L. Causey, *Unity of Science.* 1977, VIII+185 pp.
110. Richard Grandy, *Advanced Logic for Applications.* 1977 (forthcoming).

111. Robert P. McArthur, *Tense Logic*. 1976, VII + 84 pp.
112. Lars Lindahl, *Position and Change: A Study in Law and Logic*.1977, IX + 299 pp.
113. Raimo Tuomela, *Dispositions*. 1977 (forthcoming).
114. Herbert A. Simon, *Models of Discovery and Other Topics in the Methods of Science*, Boston Studies in the Philosophy of Science (ed. by Robert S. Cohen and Marx W. Wartofsky), Volume LIV. 1977 (forthcoming).
115. Roger D. Rosenkrantz, *Inference, Method and Decision*. 1977 (forthcoming).
116. Raimo Tuomela, *Human Action and Its Explanation. A Study on the Philosophical Foundations of Psychology*. 1977 (forthcoming).
117. Morris Lazerowitz, *The Language of Philosophy*, Boston Studies in the Philosophy of Science (ed. by Robert S. Cohen and Marx W. Wartofsky), Volume LV. 1977 (forthcoming).
118. Tran Duc Thao, *Origins of Language and Consciousness*, Boston Studies in the Philosophy of Science (ed. by Robert S. Cohen and Marx. W. Wartofsky), Volume LVI. 1977 (forthcoming).
119. Jerzy Pelc, *Polish Semiotic Studies, 1894–1969*. 1977 (forthcoming).
120. Ingmar Pörn, *Action Theory and Social Science. Some Formal Models*. 1977 (forthcoming).
121. Joseph Margolis, *Persons and Minds*, Boston Studies in the Philosophy of Science (ed. by Robert S. Cohen and Marx W. Wartofsky), Volume LVII. 1977 (forthcoming).

SYNTHESE HISTORICAL LIBRARY

Texts and Studies
in the History of Logic and Philosophy

Editors:

N. KRETZMANN (Cornell University)
G. NUCHELMANS (University of Leyden)
L. M. DE RIJK (University of Leyden)

1. M. T. Beonio-Brocchieri Fumagalli, *The Logic of Abelard*. Translated from the Italian. 1969, IX + 101 pp.
2. Gottfried Wilhelm Leibniz, *Philosophical Papers and Letters*. A selection translated and edited, with an introduction, by Leroy E. Loemker. 1969, XII + 736 pp.
3. Ernst Mally, *Logische Schriften*, ed. by Karl Wolf and Paul Weingartner. 1971, X + 340 pp.
4. Lewis White Beck (ed.), *Proceedings of the Third International Kant Congress*. 1972, XI + 718 pp.
5. Bernard Bolzano, *Theory of Science*, ed. by Jan Berg. 1973, XV + 398 pp.
6. J. M. E. Moravcsik (ed.), *Patterns in Plato's Thought. Papers Arising Out of the 1971 West Coast Greek Philosophy Conference*. 1973, VIII + 212 pp.
7. Nabil Shehaby, *The Propositional Logic of Avicenna: A Translation from al-Shifā: al-Qiyās*, with Introduction, Commentary and Glossary. 1973, XIII + 296 pp.
8. Desmond Paul Henry, *Commentary on De Grammatico: The Historical-Logical Dimensions of a Dialogue of St. Anselm's*. 1974, IX + 345 pp.
9. John Corcoran, *Ancient Logic and Its Modern Interpretations*. 1974, X + 208 pp.
10. E. M. Barth, *The Logic of the Articles in Traditional Philosophy*. 1974, XXVII + 533 pp.
11. Jaakko Hintikka, *Knowledge and the Known. Historical Perspectives in Epistemology*. 1974, XII + 243 pp.
12. E. J. Ashworth, *Language and Logic in the Post-Medieval Period*. 1974, XIII + 304 pp.
13. Aristotle, *The Nicomachean Ethics*. Translated with Commentaries and Glossary by Hypocrates G. Apostle. 1975, XXI + 372 pp.
14. R. M. Dancy, *Sense and Contradiction: A Study in Aristotle*. 1975, XII + 184 pp.
15. Wilbur Richard Knorr, *The Evolution of the Euclidean Elements. A Study of the Theory of Incommensurable Magnitudes and Its Significance for Early Greek Geometry*. 1975, IX + 374 pp.
16. Augustine, *De Dialectica*. Translated with Introduction and Notes by B. Darrell Jackson. 1975, XI + 151 pp.